Climate Change Action
and Cosmic Dynamics

by Rolf A. F. Witzsche

Contents

5

Ice Age – Climate Change
and the book
Climate Change Action and Cosmic Dynamics

Humanity is not a player on the Climate Change front. All Climate Change Actions are cosmic in origin. We can't affect the climate. I wish we could. It would enable us to avoid the next Ice Age. But we can't. We can only respond to the big and small climate changes that the cosmic dynamics impose. We are out of our league in the climate arena. And what we call climate science is clogged up with mysticism, such as Manmade Global Warming caused by CO_2 emissions. Manmade Global Warming is scientifically impossible, except in dreams.

CO_2 is not a climate factor. The big climate factor is cloudiness, which is controlled by changing solar cosmic-ray flux that reflects changing solar activity. And all of that is caused by the ever-changing cosmic dynamics that affect our Sun, which we have concrete measured evidence for in corresponding Carbon-14 ratios.

Ice Ages result when thresholds are crossed in cosmic dynamics. We see evidence for that in diminishing solar wind and diminishing magnetic fields. The CO_2 greenhouse effect is too minuscule to affect anything in comparison with the big Cosmic climate forcing. The CO_2 Climate Change fears are based on illusions and lies, no matter how firmly the entire world believes the illusions and lies. Science needs to be rescued from the manmade climate-change trap, for the liberation of humanity from the consequences of the illusions and lies.

Numerous fields of evidence tell us that the next Ice Age is near. That's where the truth begins. Most of the evidence was discovered in the 1990s and thereafter. Some evidence is measured in ice cores; some is measured in space, by satellites. Some measurements are also made on the ground in terms of measurements of the Earth's magnetic-pole drift observed in northern Canada. All of this is seen combined with high-energy physics experiments at a leading national laboratory, and is also explored in the small in static experiments.

Against the background of these widely diverse types of evidence that have been recently discovered, the historic Little Ice Age in the 1600s, takes on a new dimension as a yardstick for measuring the future that by this evidence promises to be up to 40-times colder than the Little Ice Age had been. It qualifies for the term, Absolute! The evidence poses a great challenge ahead. Are we ready to respond? The Ice Age phase shift in climate is a stark in differences as night and day, and similarly fast.

In the Little Ice Age between 10% and up to 30% of the populations in Europe had perished by starvation. The last Big Ice Age was evidently vastly harsher. Only 1-10 million people emerged from it alive. That's all we had after 2 million years of development. We want to do far better this time around; and we can, with large-scale technological infrastructures for our food supply. But will we create them? Will we get the job done in the 30 years that we still have left before the Ice Age starts anew? Will we even consider it? And how certain are we that the phase shift to the next glaciation period will begin, as the evidence suggests, in the 2050s? We have no slack on this front. By failing itself on this absolute front, humanity actively commits suicide.

So, what will the answer be? Will we move with the evidence? Or will we lay ourselves down to die by default?

It takes an independent researcher to brake the taboos that have kept mainstream cosmology imprisoned, increasingly, during the past century, even while what is regarded as taboo is known to be wrong.

The Illustrated Science series is intended to open the scene beyond the threshold of accepted taboos, to where the actual physical evidence speaks for itself.

The scope of the existential challenge that the Ice Age brings with it, takes astrophysics out of the academic domain and places it into the foreground as one of the most-critical issues of our time. The big Climate Change events that have already worldwide effects are mere fringe effects in the flow of the ever-changing cosmic dynamics. The big effect, when the Ice Age begins anew, promises to be caused by a dimmer and colder Sun with 70% less radiated energy. This defines our climate future.

Sure, we can live with all that by creating new platforms for agriculture that are able to operate under Ice Age conditions. But will we do it? The task is enormous. Or will we fail ourselves on this front? We have no reason to allow us to fail. We have the materials and energy resources on hand to accomplish everything that is required for us to continue to live in an Ice Age World. But will we do it? The big question that never goes away, therefore, is; will we develop our inner resources as human beings sufficiently to get the job done, and to get it done in time? Or will we do nothing, ignore the challenge, and condemn our children and one-another to an agonizing death by starvation? That's the choice.

Towards meeting the inner challenge, I have created the epic series of novels, The Lodging for the Rose. And further, towards meeting the science challenge, I have produced numerous research books and several dozen exploration videos that the Illustrated Science series is modeled after. The work is the result of a quarter century of research, for which numerous elements of evidence in related fields came to light during the timeframe of my research.

It is my hope that the work that went into all of these projects will help in some degree - for humanity that we are all a part of - to write itself a ticket to have a future.

High-resolution color images, of the images in this book, can be obtained at www.iceagetheatre.ca

*The maximum climate goal

The maximum climate goal that the United Nations Climate forum seeks in terms of preventing manmade global warming for all times to come, is already happening. It is here! Open your eyes! End the dreaming!

What science seeks in its dreaming

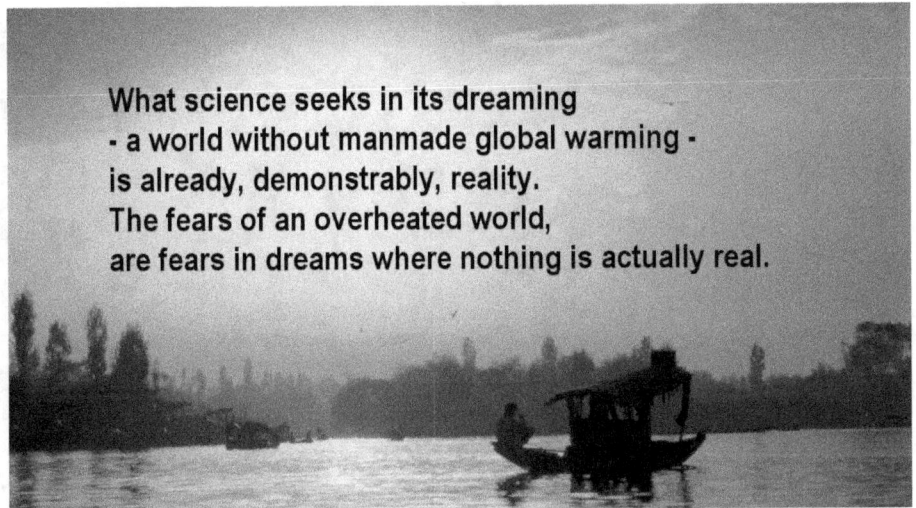

What science seeks in its dreaming - a world without manmade global warming - is already, demonstrably, reality. The fears of an overheated world, are fears in dreams where nothing is actually real.

A world of living, building, creating

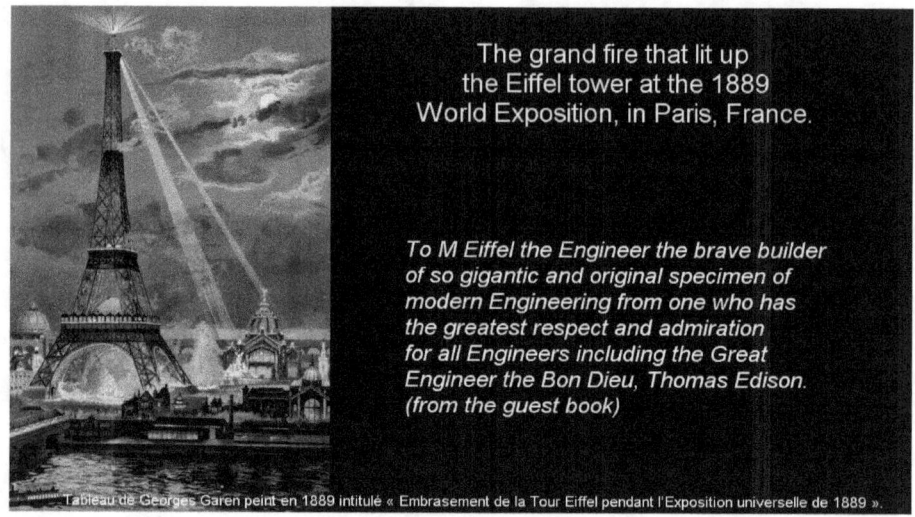

The grand fire that lit up
the Eiffel tower at the 1889
World Exposition, in Paris, France.

*To M Eiffel the Engineer the brave builder
of so gigantic and original specimen of
modern Engineering from one who has
the greatest respect and admiration
for all Engineers including the Great
Engineer the Bon Dieu, Thomas Edison.
(from the guest book)*

Tableau de Georges Garen peint en 1889 intitulé « Embrasement de la Tour Eiffel pendant l'Exposition universelle de 1889 ».

The human world is a world of living, building, creating. It is a world of beauty, art, science, music, and caring', where the most amazing is totally real and the power of the human being is demonstrated in the small and in the large, like the grand fire that lit up the Eiffel tower at the 1889 World Exposition, in Paris, France. The world was waking up in those days to the great human potential that still remains a light for all times.

We are told that optimism is archaic

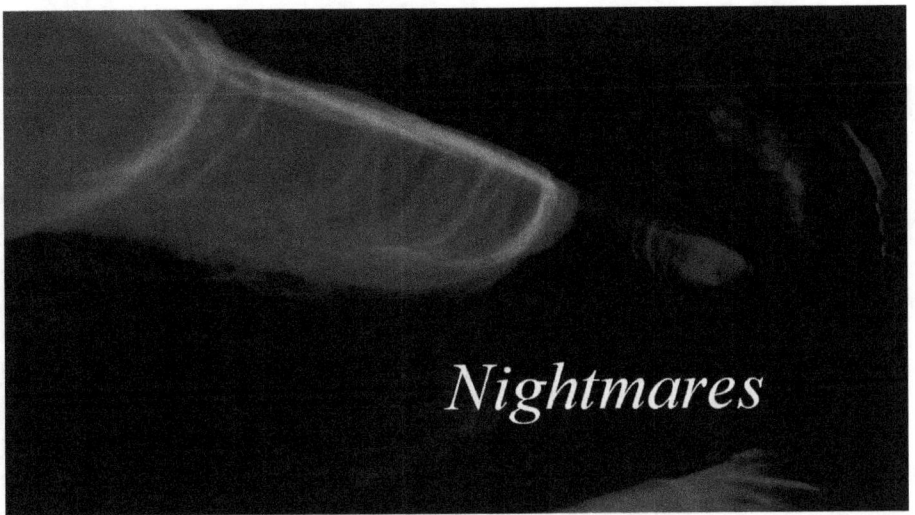

But we are told that this optimism is archaic, as many now believe in their dreams and nightmares. We are told by the elite of the world that the Earth is fast overheating by the consequences of humanity living on the Earth.

We are told that Carbon energy production must stop

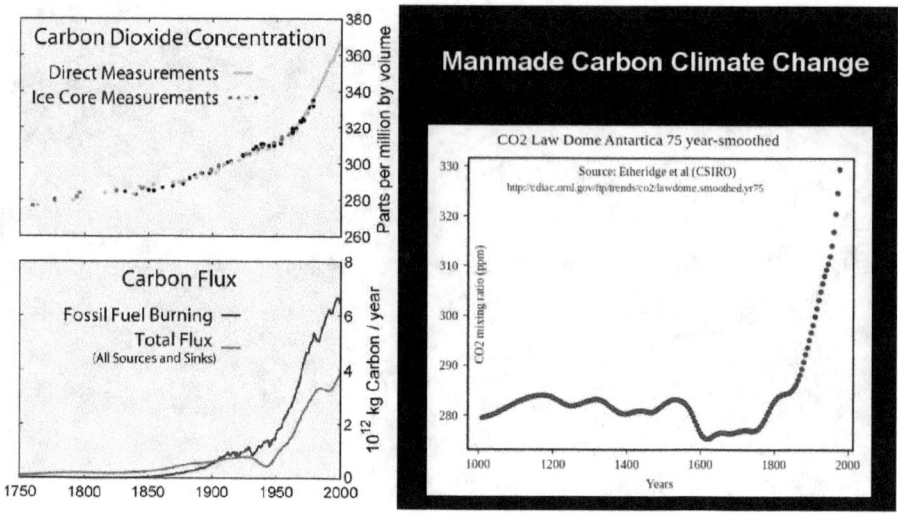

Each breath we take leaves in its wake a cloud of carbon molecules called CO2, classified a dangerous pollutant, termed a greenhouse gas, addressed with fear for is rapidly increasing.

We are told that with each house we built, each car we drive, each factory we operate, we add to the pollution that drives up the carbon global warming climate change. We are told that the pollution we emit has been skyrocketing since the dawn of industrialization. We are told that this mistake in history must be reversed. Carbon energy production must stop. The human presence must be culled back to less than a billion people. For this demand, industries are being destroyed, and the food that humanity needs to live is being burned in gigantic quantities.

Is this a dream? Or is this real?

What you will see when you awake

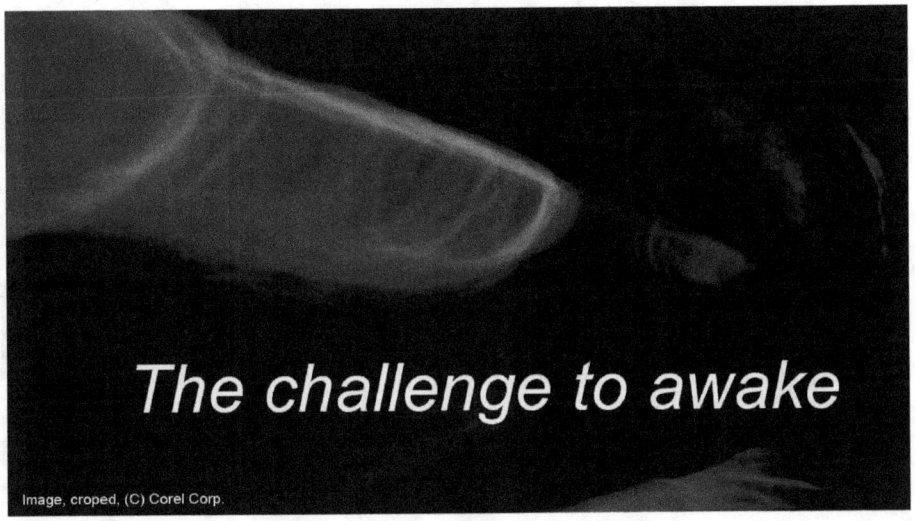

Image, croped, (C) Corel Corp.

Let me show you what you will see when you awake and stop dreaming. That's the challenge.

A world quite different than in dreams

Wikipedia - freeways

You will see a world that is quite different than in dreams. In waking to reality you discover that no form of human action, no matter how big, has the power to cause climate change on Earth.

Climate change originates with the Sun

You discover that every climate change in history, even the worst of them, originates with the Sun.

Minute forms of weakening on the Sun

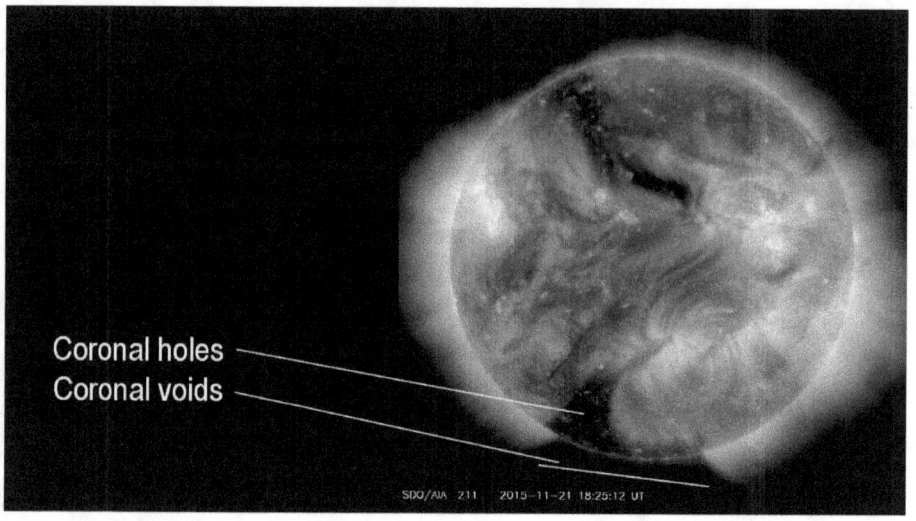

You discover that minute forms of weakening on the Sun have large affects on the cosmic-ray flux streaming from the Sun.

Solar cosmic-ray flux changes in cloudiness on Earth

ISS-34 - Stratocumulus clouds

You will discover that the invisible changes in solar cosmic-ray flux cause large changes in cloudiness on Earth, which changes the amount of the radiated energy from the Sun that is reflected back into space and thereby becomes lost to us.

Solar cosmic-ray measured in carbon-14 ratios

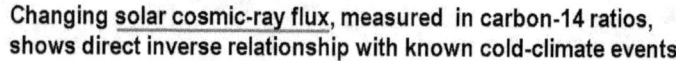

Changing **solar cosmic-ray flux**, measured in carbon-14 ratios, shows direct inverse relationship with known cold-climate events

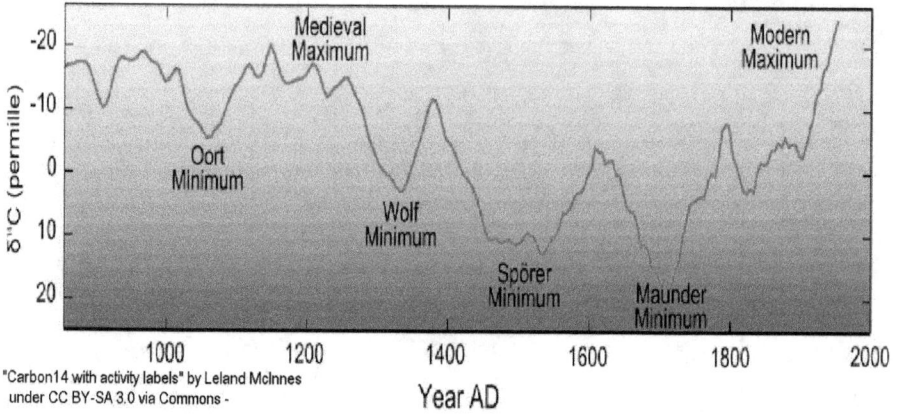

You will discover that historic solar cosmic-ray flux can be precisely measured in carbon-14 ratios, and that the measured high-flux periods correspond with cold times, and low-flux periods with warm periods.

CO2 greenhouse actions too minuscule

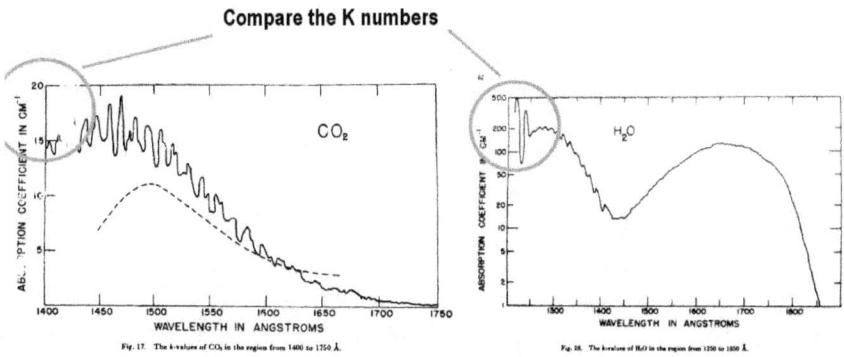

Compare the K numbers

From a 1953 study by the Geophysics Research Directorate of the Air Force Cambridge Research Center Cambridge, Massachusetts - http://www.dtic.mil/cgi-bin/GetTRDoc?AD=AD0019700

You will discover that CO2 greenhouse actions are too minuscule to cause any measurable climate effects.

Increase in CO2 is not manmade

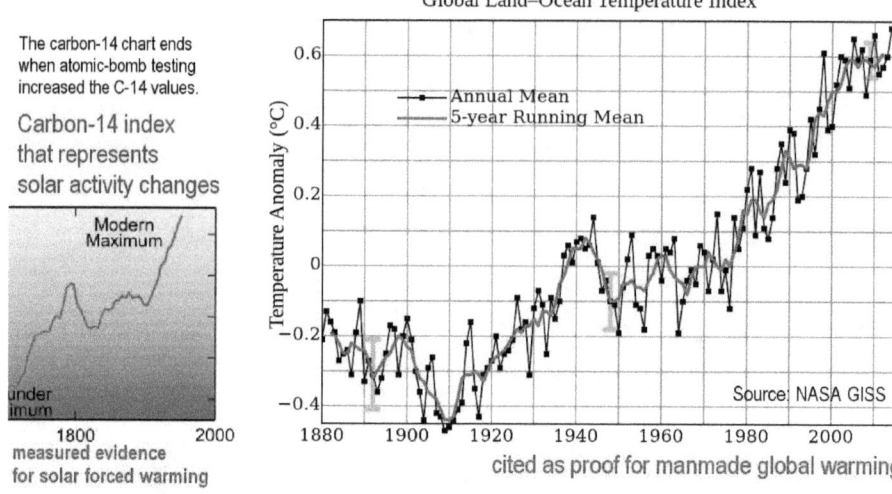

The carbon-14 chart ends when atomic-bomb testing increased the C-14 values.

Carbon-14 index that represents solar activity changes

Modern Maximum

under imum

1800   2000
measured evidence for solar forced warming

Global Land–Ocean Temperature Index

Temperature Anomaly (°C)

Annual Mean
5-year Running Mean

Source: NASA GISS

1880   1900   1920   1940   1960   1980   2000

cited as proof for manmade global warming

You will discover that the measured large increase in CO2 since the dawn of industrialization is not manmade.

The global CO2 recycling conveyor system

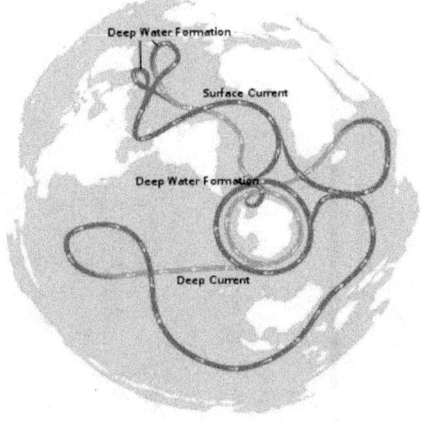

The ocean currents
conveyor belt
centered on the
deep cold waters
encircling Antarctica

"Conveyor belt" by Avsa - under CC BY-SA 3.0 via Commons -

You will discover that the global CO2 recycling conveyor system
gives us back today, by its long transit time, the high volume of CO2
that had been dissolved in the oceans during the Little Ice Age and
before.

The next Ice Age in 30 years

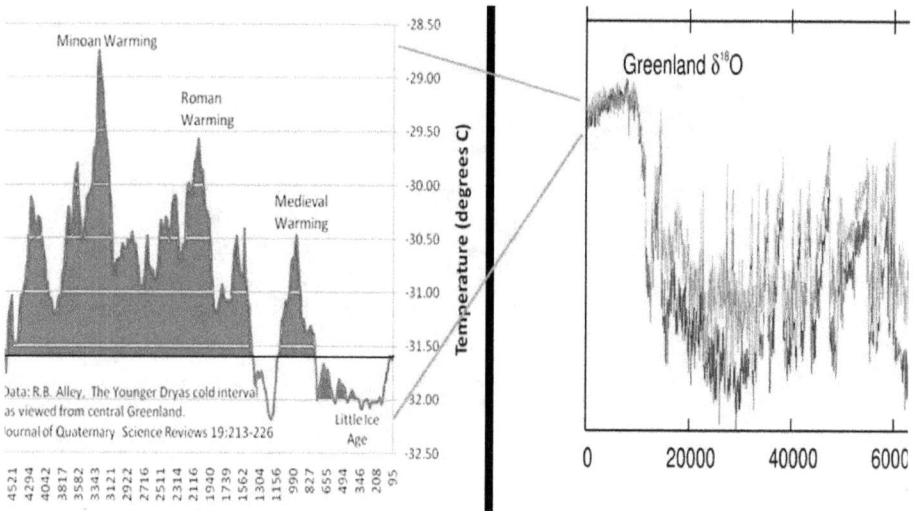

You will discover that you have no reason to fear any manmade climate changes, but that you would wish instead that we had this capability as a means to avoid the next Ice Age in 30 years, that's the only climate factor with an existential challenge attached.

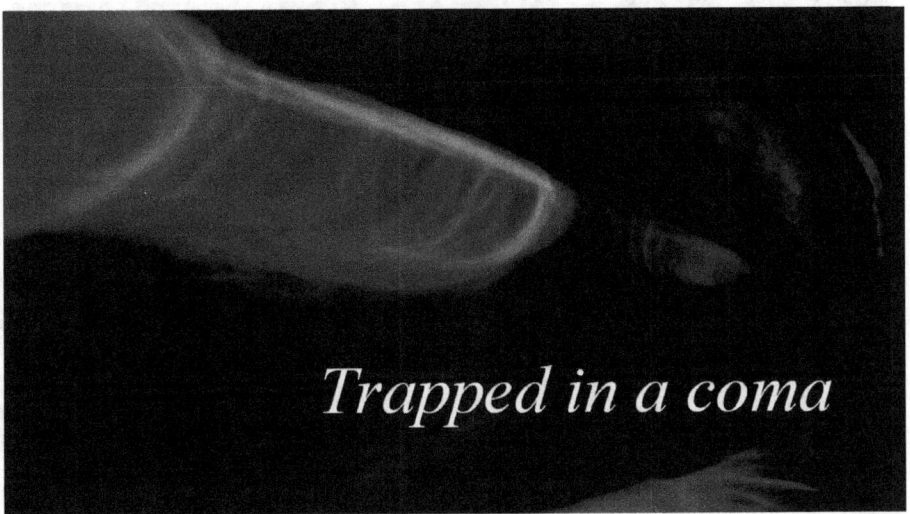

Trapped in a coma

The reason why the world is in a coma today, and is largely unaware of the reality that actual science unfolds, ironically appears to be rooted in a scientific discovery itself, that takes us back in time to the Roman era.

Quintus Fabius Maximus

The Roman politician and military general

Quintus Fabius Maximus

Founder of the principle of the Strategic Defence (280 BC - 203 BC) which became inverted into Modern Irregular Warfare

"The Shield of Rome"

Founder of the principle of the Strategic Defence (280 BC - 203 BC) which became inverted into Modern Irregular Warfare.

Early in the Roman era, the Roman politician and military general, Quintus Fabius Maximus, had pioneered a principle for the protection of his country, which might be termed the principle of the strategic defence.

Rome had been under attack by the vastly superior forces, led by the Punic Carthaginian military commander, Hannibal Barca, who is considered one of the greatest military commanders in history. Hannibal had occupied much of Italy for 15 years, but had never achieved a complete victory. Fabius Maximus had developed a strategy of irregular warfare that had enabled the small Roman forces to prevent the superior opponent from defeating the state. The method that he applied is simple.

His strategic defence was, not to attack Hannibal, but to grind the opposing forces into dust by denying them the means to operate. He attacked Hannibal's logistics. Quintus Fabius became famous for it. He was honoured for his strategy of irregular warfare that saved

Rome, not by direct confrontation, but by grinding the superior opponent down, and down, into the dust, for which he earned the title, "The Shield of Rome."

A Phase Shift was Staged

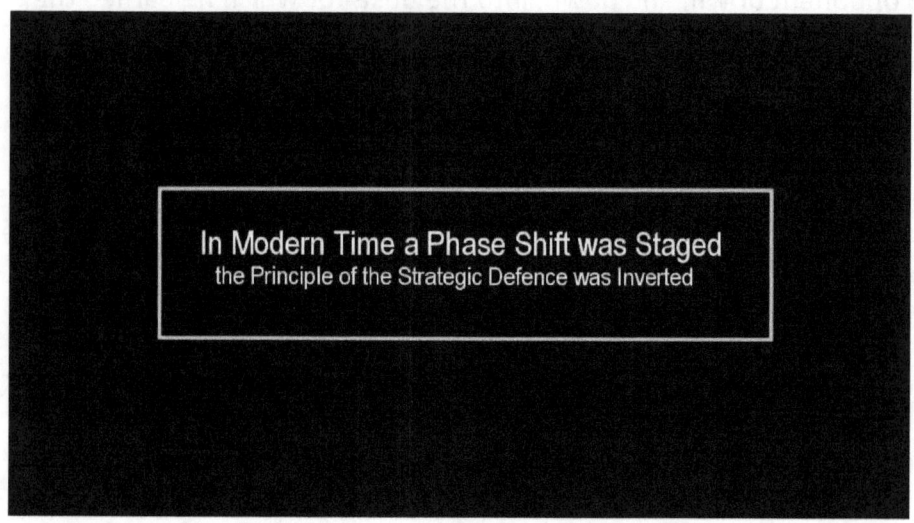

In Modern Time a Phase Shift was Staged. The Principle of the Strategic Defence was Inverted.

Fabius Maximus was modernized

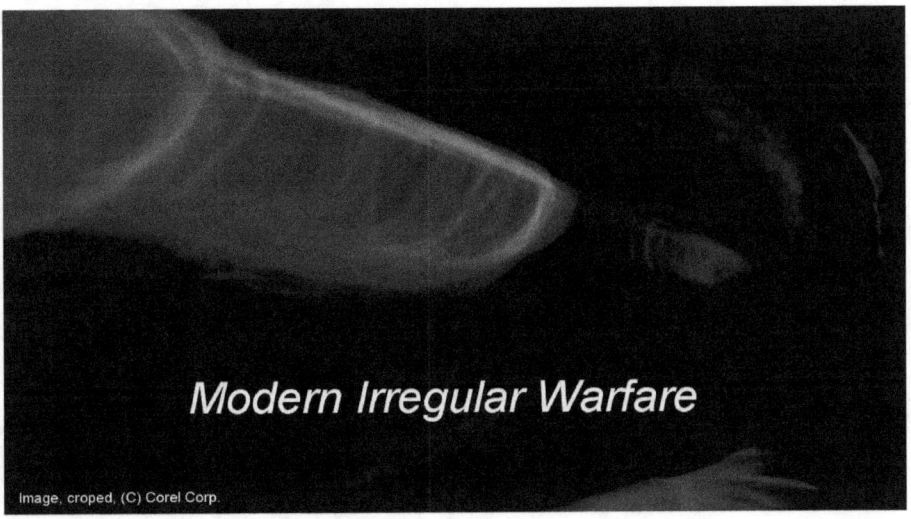

Modern Irregular Warfare

Image, croped, (C) Corel Corp.

The pioneered strategy of Quintus Fabius Maximus, was not forgotten. It was modernized. It was transformed from a platform for the strategic defence of a nation, into a comprehensive informal system of modern irregular warfare that operates in ambush to maintain an empire against the scientific and moral power of an advancing humanity.

Human progress a threat to empire

The principle of modern irregular warfare appears to have been chosen by the masters of the oligarchic system of empire, for the defence of their system that lacks itself a foundation to exist, but hangs on to its power against the vastly superior force of human progress that stands as a threat to the system of empire. The masters fear the power of a humanist cultural renaissance and advancing scientific, technological, and industrial development that creates a higher-level progressive world in which the system of empire would have no place to exist.

The grind-down method of Fabius Maximus

Partners in a common cause?

Quintus Fabius Maximus

COP15
COPENHAGEN
UN CLIMATE CHANGE CONFERENCE 2009

COP15 official logo

It would be surprising in the context of the critical strategic situation that the masters of empire find themselves in evermore, if the grind-down method of general Quintus Fabius Maximus would not be applied again as an emergency platform to maintain the decaying system of oligarchic control against the vastly superior potential of an advancing humanity. The end-result of this process became the principle of Modern Irregular Warfare, as it is now called.

It would also be surprising if the now ongoing war against humanity and human progress, would not include a component of war that is directed against science and truth, which are the very foundation of advancing civilization. And it would be further surprising if the manmade carbon climate change project, which has not a single item of scientific evidence standing in support of it, would not have been created as a card in the deck of modern irregular warfare to maintain the system of empire that has no basis to exist in a progressive human world, but still continues to wield its historic influence and political and financial control.

*What is the evidence telling us?

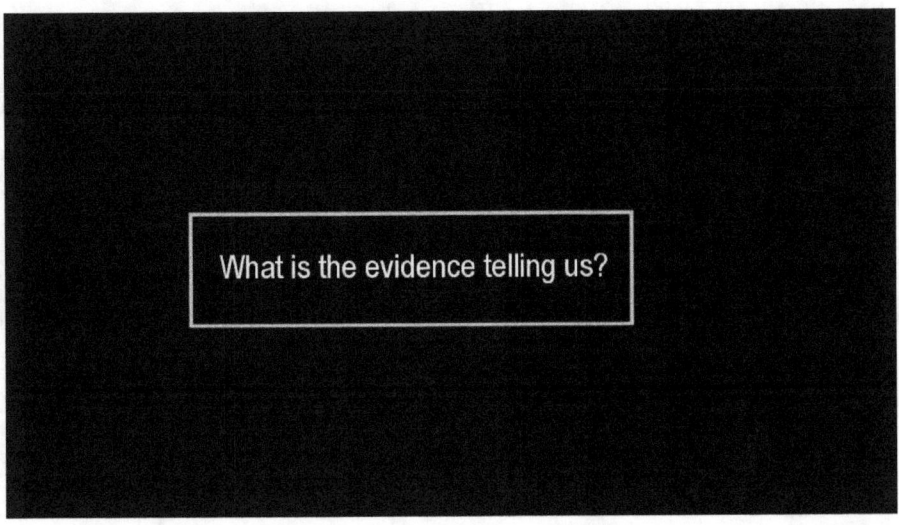

** What is the evidence telling us?

Global thermonuclear war would

Annihilation is assured

500,000 times
Hiroshima
in one hour

Castle Bravo - the first U.S. test of a dry fuel thermonuclear hydrogen bomb - March 1, 1954 at Bikini Atoll, Marshall Islands

The evidence is sometimes surprising. While it is possible that the worst of human action in the form of a global thermonuclear war would cause a prolonged nuclear winter by injecting vast amounts of material into the atmosphere and even the stratosphere, which would wipe out food production around the world and starve the survivors to death, the giant climate upset would nevertheless remain to be but a temporary thing, perhaps of a few decades in duration, which, of course, only a few people might survive, if any. But this isn't what the climate hoopla is about, is it?

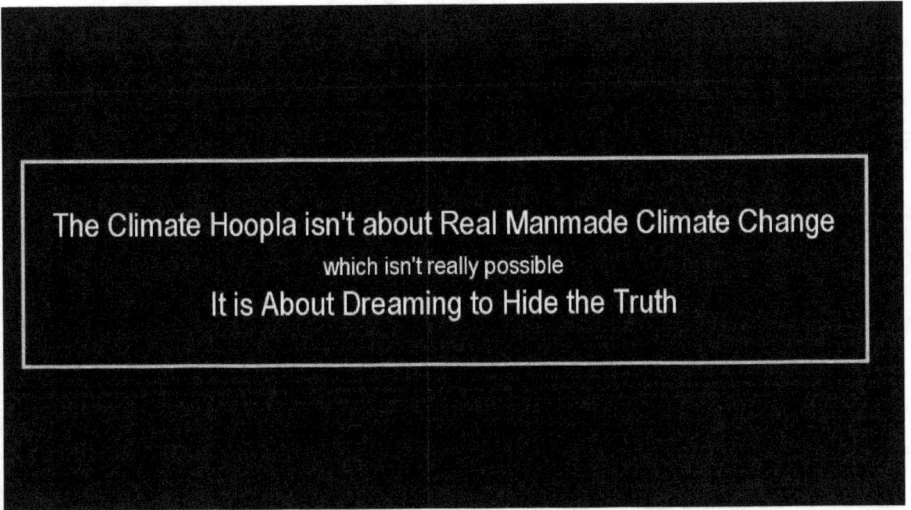

** The Climate Hoopla isn't about Real Manmade Climate Change, which isn't really possible. It is About Dreaming to Hide the Truth.

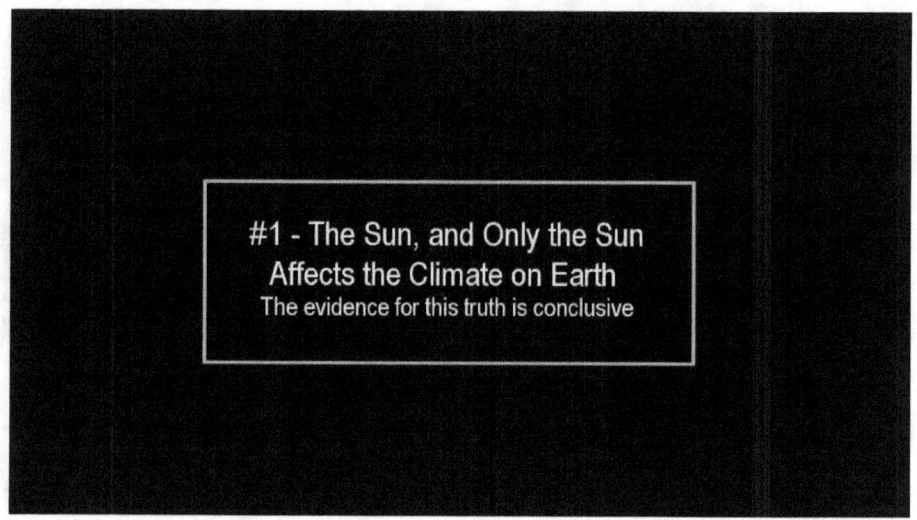

**#1 - The Sun, and Only the Sun affects the Climate on Earth. The evidence for this truth is conclusive.

Climate change trap has no basis

The manmade climate change trap has no basis. The dreaming is focused on minute changes in the atmosphere, extended over long periods, that are deemed accumulative and increasing, which are said to have raised the density of carbon gases in the air that are said to threaten the world with run-away global warming, melt the ice caps, and flood large areas of low-elevation land.

This tragic, long-term prediction is false, though it has been erroneously accepted for a long time, and has caused increasingly, enormous consequences in human living on an unimaginable scale, while no measurable global warming has actually occurred.

Does this ring like a paradox? Let's try to solve the paradox by searching out what is actually real.

Historic climate records in ice samples

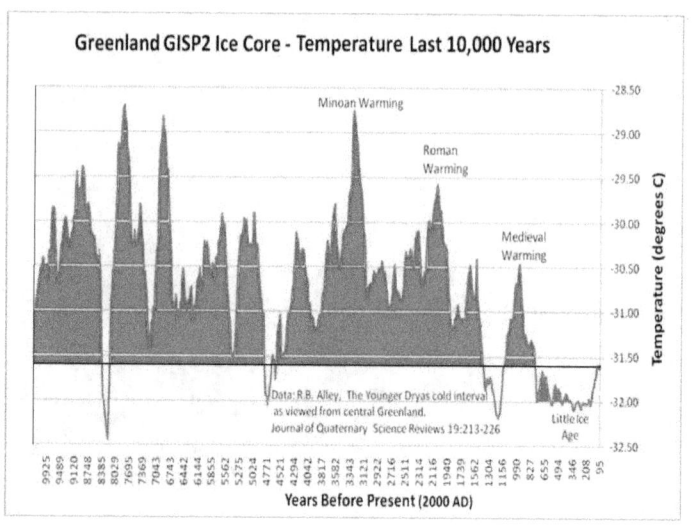

When we look at historic climate records in ice samples from Greenland, going back in time through the entire period where what we call civilization has developed, we recognize that some extremely long climate changes have occurred on our planet. But those, according to all evidence, have been the result of cosmic actions, not human actions. The human presence was minuscule for most of the long span of human history, almost to the present time. So, what did cause these very large and sometimes rapid climate changes that we have records of in ice core samples as shown here for the last 10,000 years?

There is only one cause possible, namely that the numerous large climate fluctuations that we have evidence of, are expressions of cosmic fluctuations.

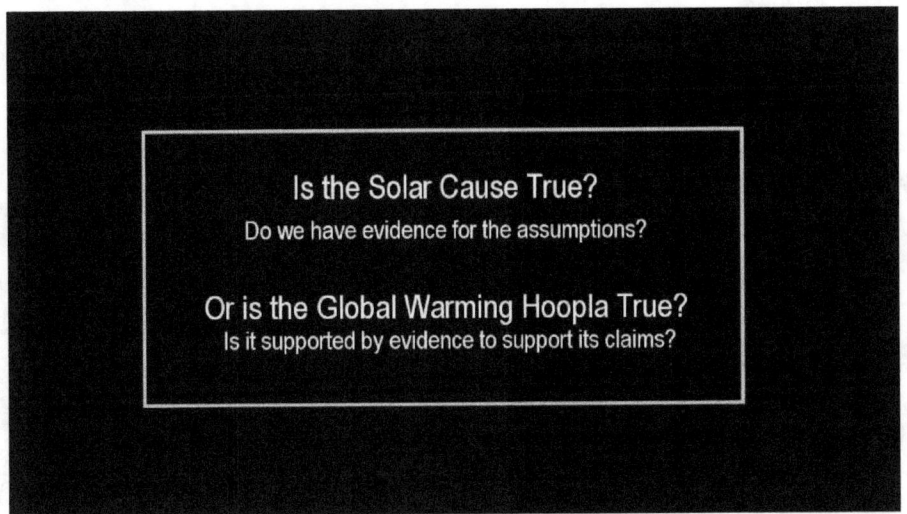

** Is the Solar Cause True? Do we have evidence for the assumptions? Or is the Global Warming Hoopla True? Is it supported by evidence to support its claims?

The fact is, the global warming hoopla has no evidence for its claim. The Earth has been in a cooling trend since 1998, even while the CO_2 levels have been rising. So, let's look at the cosmic scene as a cause for the climate changes. What evidence do we find there?

When we look for cosmic factors

Well, when we look for cosmic factors, the Sun, which creates our climate in the first place, is the prime cosmic factor for all the climate changes on Earth that have occurred.

The Sun does produce cosmic-ray flux

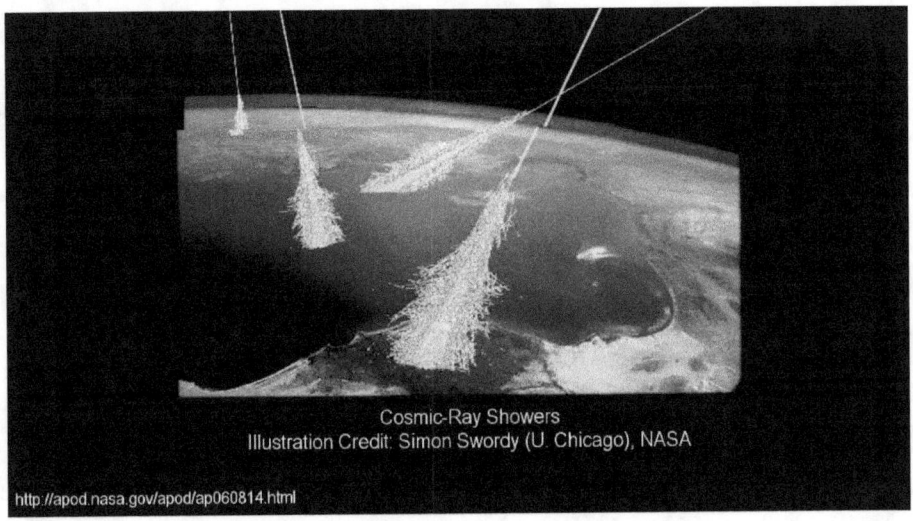

Cosmic-Ray Showers
Illustration Credit: Simon Swordy (U. Chicago), NASA

http://apod.nasa.gov/apod/ap060814.html

The Sun does produce cosmic-ray flux. We can measure it. We have measured it with satellites, and we have measured its changing.

Changes in solar cosmic-ray flux

ISS-34 - Stratocumulus clouds

We also know that changes in solar cosmic-ray flux have a big impact on the climate on Earth, by their effect on cloudiness.

Solar cosmic-ray fluctuations

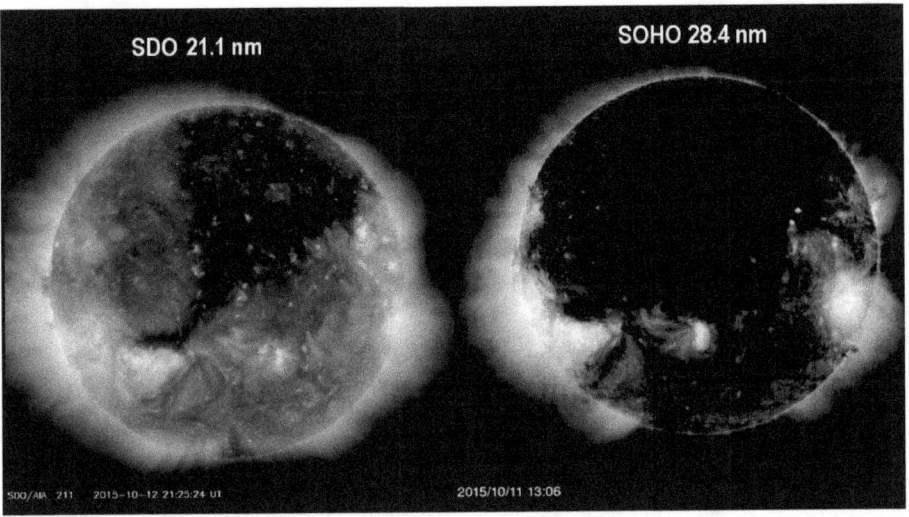

The solar cosmic-ray fluctuations, in turn, are themselves demonstrably caused by cosmic factors outside of the solar system that affect our sun in ways that allow coronal holes to occur. This cosmic linkage has always existed. This is the truth.

Cosmic actions on our climate

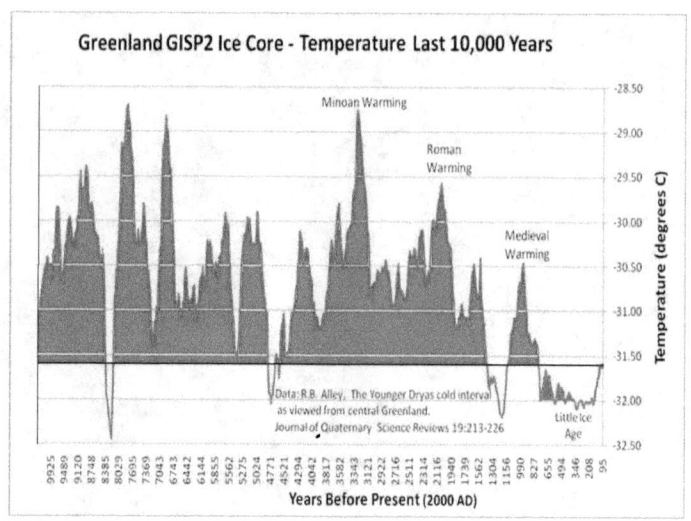

Some of the cosmic actions on our climate have been enormous, as you can see. They have caused three great warm periods in the last 4,000 years. Some are so distant and large that they are no longer imaginable in the present for the lack of comparable experiences. The most dramatic period of global warming in recent times, which has been experienced in part, has occurred at the end of the Little Ice Age, from the 1850s on. This warming period in recovering from the Little Ice Age coincides roughly with the time of the start of the industrial revolution.

The Great Global Warming period

Glacial line at the end of the Little Ice Age

"Morteratsch- und Persgletscher" by Günter Seggebäing - Archiv Günter Seggebäing. CC BY-SA 3.0 via Commons.

The massive glacial deposits left over from the 300-years long Little Ice Age, began to melt during what became referred to in modern time as the Great Global Warming period.

From the end of the Little Ice Age to the present, carbon based energy production gave the world's atmospheric CO2 component a boost. In this time frame the atmospheric CO2 level increased from 270 parts per million, which some dispute, to 390 parts per million, which is the currently measured average.

Let's assume that the 44% increase really did happen, which is likely. Is it reasonable to assume then, that the experienced re-warming of the Earth after the Little Ice Age, has resulted from the increase in atmospheric CO2, and that the entire increase of it was manmade, simply because it occurred during the industrial revolution? The entire global warming hoopla is based on that.

Or was the re-warming of the Earth a natural consequence of cosmic factors that had nothing to do with human activity and CO2 climate forcing, so that both factors simply became abused in retrospect for a scare campaign for political objectives, without

anything real standing behind it?
How much truth is in that? In what direction does the evidence point?

We have to offer up scary scenarios

A dozen years of lies

Lord Christopher Monckton exposes
the climate fraud on the world stage.

thousands of leaked e-mails w. evidence
of fraud, falsified 'official' data
destruction of historic data

computer division at the Climate Research
reveals programs and data in a hopeless,
tangled mess so that global temperature
trends were simply made up.

COP15
COPENHAGEN
UN CLIMATE CHANGE CONFERENCE 2009

Global warming proponent Stephen Schneider: "...[W]e
have to offer up scary scenarios, make simplified, dramatic
statements, and make little mention of any doubts we may
have. Each of us has to decide what is the right balance between being effective and being honest."

Nov. 1997 - 21st Century Science and Technolgy

Lies exposed at the U.N World Conference

One of the activists of the political climate campaign has answered the question in this way, saying, "We have to offer up scary scenarios, make simplified dramatic statements, and make little mention of any doubts we may have. Each of us has to decide himself what is the right balance between being effective and being honest."

That this direction became the general trend, came to light years later near the time of the 2009 U.N. Climate Conference in Copenhagen, where a mass of leaked e-mails from a leading climate institution had revealed that the much heralded climate data that the scare campaign was based on, was simply made up.

While this is all interesting, it really doesn't prove anything.

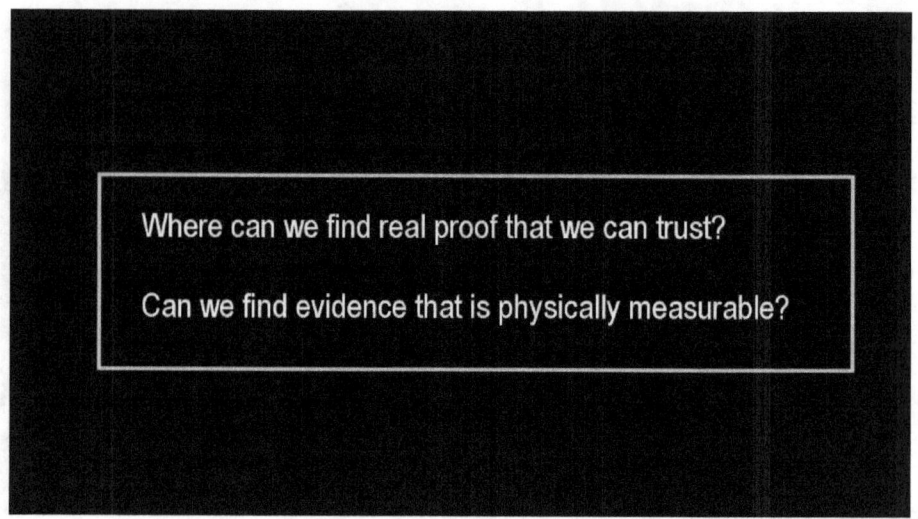

** Where can we find real proof that we can trust? Can we find evidence that is physically measurable?

Increase in CO2 is real

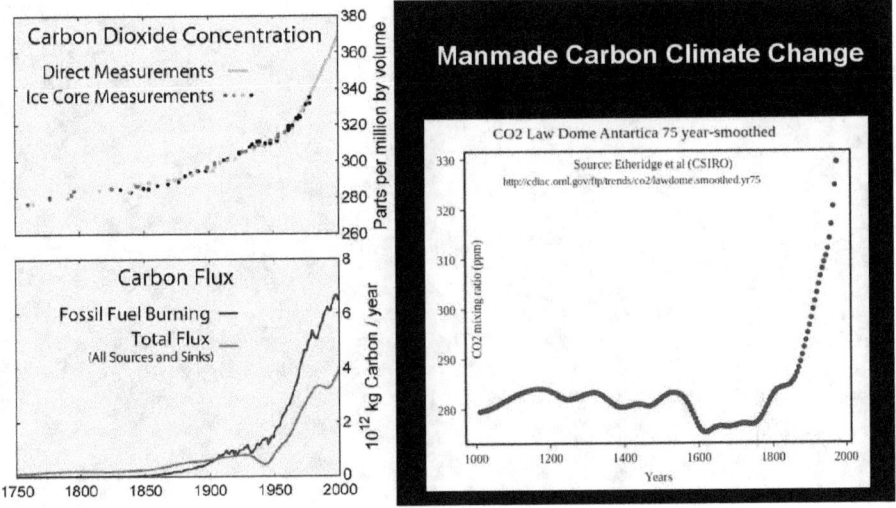

The increase in CO2 levels in the atmosphere is measurable. It has been measured. It is real.

Does this also mean that the horrific climate scare scenarios that are based on the CO2 increase, that have been promoted for the last 40 years, are also real?

The scare scenarios are promoted by the U.N. and other world organizations, and they continue to be promoted to the present. The promotion is based on the physical fact that the carbon dioxide gas, called CO2 by its chemical formula, has the capacity to absorb radiated solar energy in a process similar to light absorption, and is able to radiate the absorbed energy back into the atmosphere in a scattered fashion. That's true. CO2 does have the capacity to absorb light energy and emit it back. But is everything true?

The CO2 Greenhouse Effect

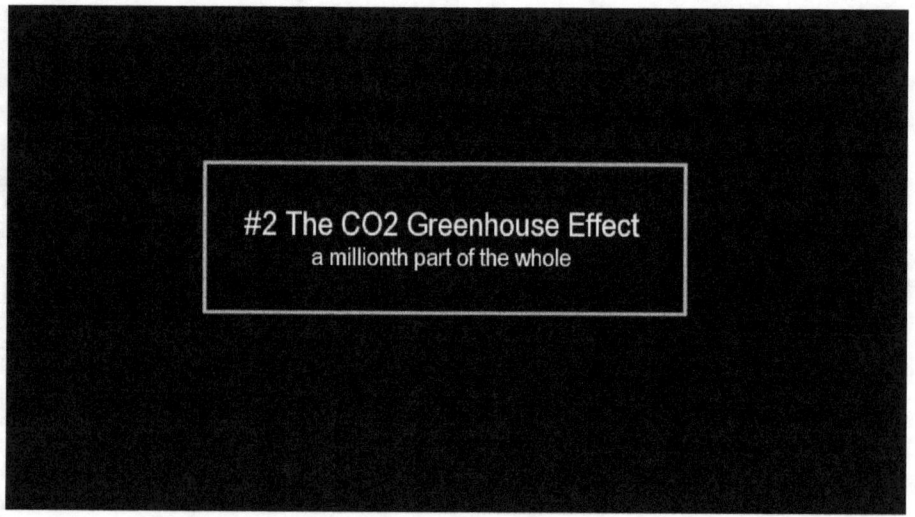

**#2 The CO2 Greenhouse Effect - a millionth part of the whole.

Energy-absorbing quality of the CO2 gas

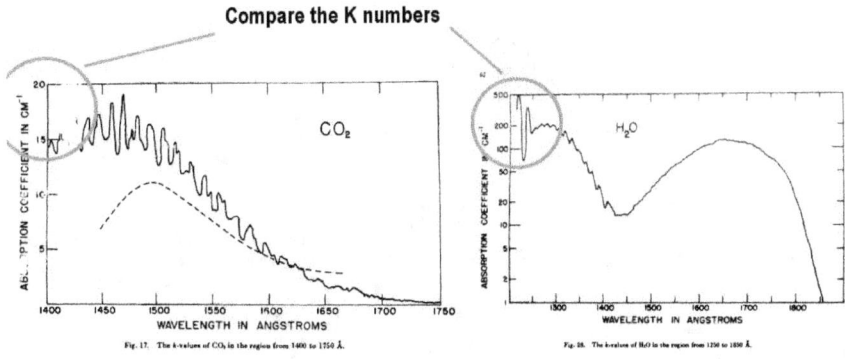

Compare the K numbers

Fig. 17. The k-values of CO₂ in the region from 1400 to 1750 Å.

Fig. 18. The k-values of H₂O in the region from 1250 to 1850 Å.

From a 1953 study by the Geophysics Research Directorate of the Air Force Cambridge Research Center Cambridge, Massachusetts - http://www.dtic.mil/cgi-bin/GetTRDoc?AD=AD0019700

On the surface it seems that manmade global warming is true. The CO_2 has increased. The energy-absorbing quality of the CO_2 gas has been measured. It has been rightfully termed a greenhouse gas. Its energy absorbing quality has also been measured. It has been measured and plotted in terms of its absorption coefficient. The coefficient has been measured across the entire light spectrum and beyond.

Here it gets interesting.

Water vapor dominates the stage!

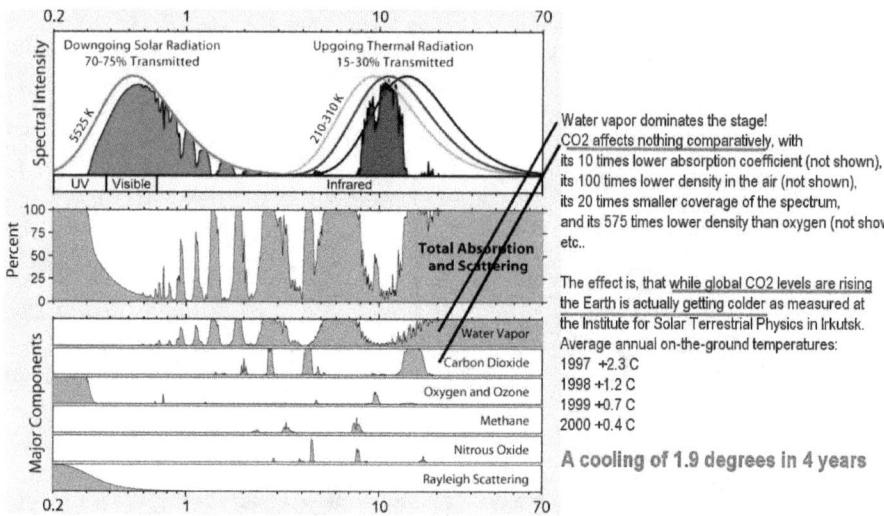

Water vapor dominates the stage!
CO2 affects nothing comparatively, with
its 10 times lower absorption coefficient (not shown),
its 100 times lower density in the air (not shown),
its 20 times smaller coverage of the spectrum,
and its 575 times lower density than oxygen (not shown),
etc..

The effect is, that while global CO2 levels are rising
the Earth is actually getting colder as measured at
the Institute for Solar Terrestrial Physics in Irkutsk.
Average annual on-the-ground temperatures:
1997 +2.3 C
1998 +1.2 C
1999 +0.7 C
2000 +0.4 C

A cooling of 1.9 degrees in 4 years

Water vapor dominates the stage!
CO2 affects nothing, comparatively, because of its 10 times lower absorption coefficient, its 100 times lower density in the air, its 20 times smaller coverage of the spectrum, its 575 times lower density than oxygen, and its effectiveness in only the low-energy end of the spectrum. The proof is in the fact that while global CO2 levels are rising the Earth is actually getting colder.

The Institute for Solar Terrestrial Physics in Irkutsk has measured a whopping 1.9 degree drop in average annual on-the-ground temperature measurements.

These facts are being ignored in the scare-hoopla theatres. Colder temperatures have been experienced all over the world, while the CO2 levels keep going up. The key for solving this paradox is located in the effect of water vapor.

The CO2 factor, in comparison

Compare the K values: 10 - 30 for CO2 vs. 100-600 for water vapor (H2O)

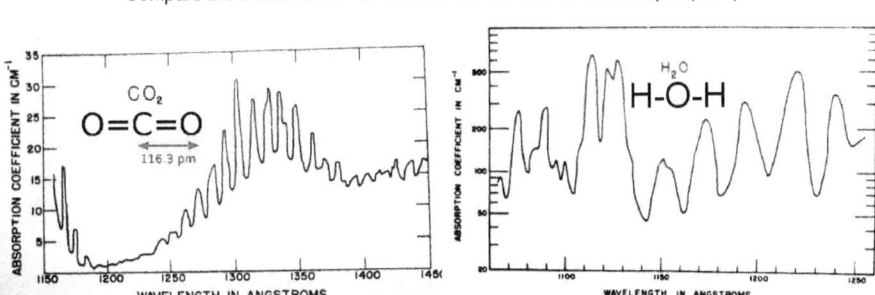

Fig. 18. The k-values of CO₂ in the region from 1150 to 1450 Å.

From a 1953 study by the Geophysics Research Directorate of the Air Force Cambridge Research Center Cambridge, Massachusetts - http://www.dtic.mil/cgi-bin/GetTRDoc?AD=AD0019700

The dramatic difference between the absorption coefficient for CO2 and for water vapor, is the result of the CO2's strong molecular bond in comparison with the extremely weak bond for the water molecule that responds more readily and on a much wider range of light energy. It would be amazing if the CO2 molecule, with this built-in deficiency, would have a significant effect on our climate, especially when one considers that water vapor, that is ten times more efficient, is typically 100 times more densely present.
In addition to all that, the giant factor that water vapor is, has its density directly affected by cosmic factors that affect the cloud-forming process. The cloud-forming process draws on the water vapor in the air, which reduces the prevailing water-vapor density in the atmosphere. With all this considered, the water vapor factor, which is directly affected by solar cosmic-ray flux density fluctuations, adds up to being a significant factor in affecting the climate on Earth.
The CO2 factor, in comparison, is so small that it is actually not

physically measurable in the real world.

It should be seen as a paradox

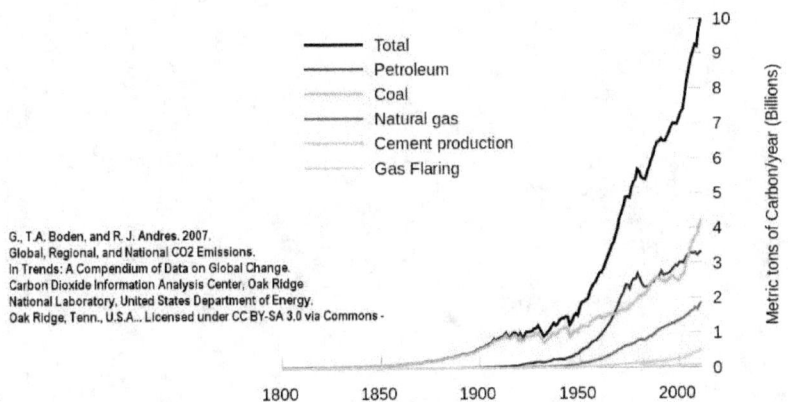

It should be seen as a paradox, what we behold in the real world. We are injecting large volumes of CO2 into the atmosphere. According to the carbon climate change models, the Earth should be roasting. Instead it is getting colder. Nor does this paradox deter the global warming song. The global warming song continues in spite of the evidence, which adds another paradox.

Lord Russell explained the paradox

Bertrand Arthur William Russell,
3rd Earl Russell, 1872 - 1970
1950 Nobel Laureate in Literature

Wikipedia

"Bad times you say, are exceptional, and can be dealt with by exceptional methods. This has been more or less true during the honeymoon period of industrialism, but it will not remain true unless the increase of population of the world is enormously diminished... War, so far, has had no very great effect on this increase, which continued through each of the world wars. [War] has been disappointing in this respect... but perhaps, biological war may prove more effective. If a Black Death could spread through the world once in every generation, survivors could procreate freely without making the world too full... The state of affairs may be somewhat unpleasant, but what of it? Really high-minded people are indifferent to happiness, especially other people's."

From *The Impact of Science Upon Society* (New York: Simon and Schuster, 1953) pp. 102-104

Lord Russell explained the paradox. As the intellectual master of the system of empire, he understood Quintus Fabius Maximus and his principle of grinding an opponent into dust by denying the opponent the logistics that are essential for a human society, or a human military force, to function. The principle was understood by the German poet Friedrich Schiller who had aimed to inspire society to rebuild itself from within, in order to free itself from the scourge of empire that had dominated Europe at the time. The idea resonated in Russia and inspired the defence of Russia against the immensely greater force of Napoleon in the War of 1812, on the principle of grinding Napoleon's force into dust by defeating its logistics. Lord Russell understood that the modern logistical base of humanity is its science and scientific and technological progress. In his book, The Impact of Science on Society, he made it rather plain that for empire to defend itself against humanity, which he defined as the enemy, humanity's modern logistical base, which is science as the keystone in human culture, must be defeated. For this he champions the mass elimination of humanity. He says in the book that wars are insufficient to have this effect. He says that the depopulation must be accomplished through the back door by waging war on science and truth.

The point is, that there is no point in one trying to understand the carbon climate issue as an issue outside the context of "modern irregular warfare." The prime objective of empire is to defend its existence. Lord Russell made it clear that humanity is the declared enemy and that science and truth is its power. This means that the carbon climate fraud, which makes sense only in the context of empire aiming to defend itself against humanity, is not about to be called off. "Modern Irregular Warfare," is a part of this strategic landscape. It is here to stay for as long as possible. When it fails, empire falls.

Modern irregular warfare more devastating on society

Dresden 1945

The concept of "modern irregular warfare," is rather simple when it is seen in the context of the oligarchic system aiming to achieve world domination without the use of world wars, that in the nuclear age are no longer possible. This means that the goal of world domination must be pursued below the threshold of world wars, by means of subversive warfare from within. During the big world wars, the goal has been to destroy the populations, their infrastructures, and their industries. This goal is now pursued subversively. The process is inexpensive, and it is vastly more destructive than open warfare had been.

It has become recognized that it is vastly more destructive and devastating on society, for it to become subjected to numerous false idealisms that render the human being as immoral, despicable, greedy, irresponsible, and worthless scum - even a danger to itself, to its resources, to its environment, and to its very existence. The depopulation card, that empire plays, fits into this deck of objectives, as does the terrorizing environmental card, and the carbon climate card. Each one of these cards is subversively destructive on society by design. Each card is expertly played.

The objective is to kill the soul

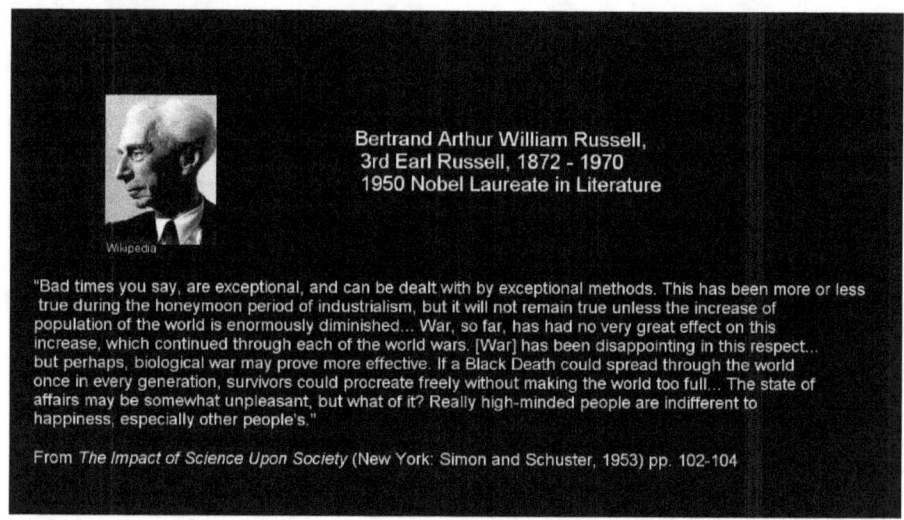

Bertrand Arthur William Russell,
3rd Earl Russell, 1872 - 1970
1950 Nobel Laureate in Literature

Wikipedia

"Bad times you say, are exceptional, and can be dealt with by exceptional methods. This has been more or less true during the honeymoon period of industrialism, but it will not remain true unless the increase of population of the world is enormously diminished... War, so far, has had no very great effect on this increase, which continued through each of the world wars. [War] has been disappointing in this respect... but perhaps, biological war may prove more effective. If a Black Death could spread through the world once in every generation, survivors could procreate freely without making the world too full... The state of affairs may be somewhat unpleasant, but what of it? Really high-minded people are indifferent to happiness, especially other people's."

From *The Impact of Science Upon Society* (New York: Simon and Schuster, 1953) pp. 102-104

Against this background the carbon scare card is not surprising. When Bertrand Russell of the British Empire system, argued for the mass killing of society with vectored diseases for the purpose of depopulation, he evidently didn't have the actual mass-murdering of society in mind. It makes no sense for the masters to kill the slaves that supply their living. The textbook objective in modern irregular warfare is far-more devastating on society than the physical murdering would be. The objective is to kill the soul: to so degrade the status of man so deeply that society becomes ashamed of itself for being alive. The endlessly-repeated depopulation song has this effect. The carbon climate change song has the same effect. They are cards of a larger deck of cards. They are in the deck side by side with the card of naked terrorism, and the ultimate-terror card of thermonuclear war. These cards are all expertly played.

Carbon climate change is a devastating card

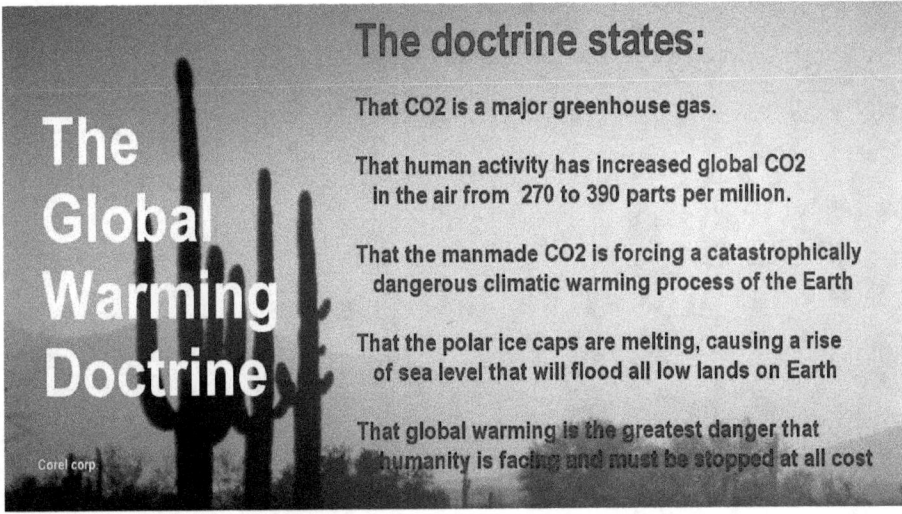

The Global Warming Doctrine

The doctrine states:

That CO2 is a major greenhouse gas.

That human activity has increased global CO2 in the air from 270 to 390 parts per million.

That the manmade CO2 is forcing a catastrophically dangerous climatic warming process of the Earth

That the polar ice caps are melting, causing a rise of sea level that will flood all low lands on Earth

That global warming is the greatest danger that humanity is facing and must be stopped at all cost

Corel corp.

The carbon climate change ideology is a devastating card that is not based on anything real. It is a card that puts a devastating problem before society for which no solution is possible in the framework in which it is presented, other than for society to commit economic suicide or to depopulate itself, since society itself is said to be the cause of the problem.
A project of this type has a devastating effect on the human mind. The mind is trapped thereby, into a self-perception for which no acceptable solution is deemed possible. In such a case, the mind reverts backwards to a more infantile state of thinking. This appears to be intentional. At the more infantile state society is more easily controlled.

Global Warming a terror phrase

The term, Global Warming, has in this context been employed as a
terror phrase that is effective, even if there is no truth behind it.
Society has become inspired by its terror song, deep in its soul, the
be ashamed to be alive as a burden to the Earth. The effect may be
the reason why the Global Warming doctrine is so intensively
guarded against the truth. To even think the truth, is slandered as
an act of Global Warming Denial. It is deemed a crime thereby to be
honest with oneself.

Depopulation card and the Global Warming card

Lord Christopher Walter Monckton in Washington, D.C.
Head of the Policy Unit United Kingdom Independence Party
Photo © 2009 Joanne Nova (CC-BY-SA) Wikipedia

Global Warming Fraud exposed by Viscount Lord Christopher Monckton

COP15
COPENHAGEN
UN CLIMATE CHANGE CONFERENCE 2009

Typically, the Depopulation card and the Global Warming card are played together to maximise their effects. Placed into skilful hands, these cards are being played from behind dense smokescreens so that the objectives are rarely ever recognized.

While in the case of the 2009 Climate Conference in Copenhagen, the climate fraud became proudly exposed by the aristocrat Lord Monckton - who cited that key climate facts were simply made up in high-level institutions, as had been revealed by leaked e-mails - he failed to expose the purpose for the fraud. He failed to see what stands behind this type of scene, behind the official face. He failed to see the connection of the fraud as an attack on humanity. He didn't mention the depopulation target, which at this time had been openly set at 2.3 billion people as the allowable optimum world population. The figure has since been reduced to less than a billion people.

The carbon climate-scare fraud

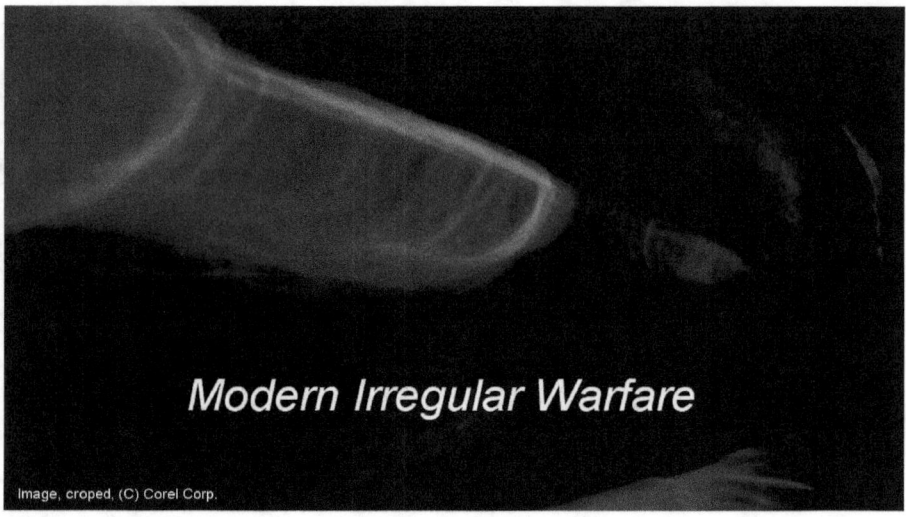

Modern Irregular Warfare

Image, croped, (C) Corel Corp.

The deployment of the Global Warming card in conjunction with the Depopulation card, signifies that the carbon climate-scare fraud is promoted as an element of the modern irregular warfare objective that the masters of empire had been playing already from the late 1700s on, though evermore vigorously now, as a means for protecting the essentially feudal platform that all systems of empire depend on.

The 40-year Global Warming irregular warfare campaign

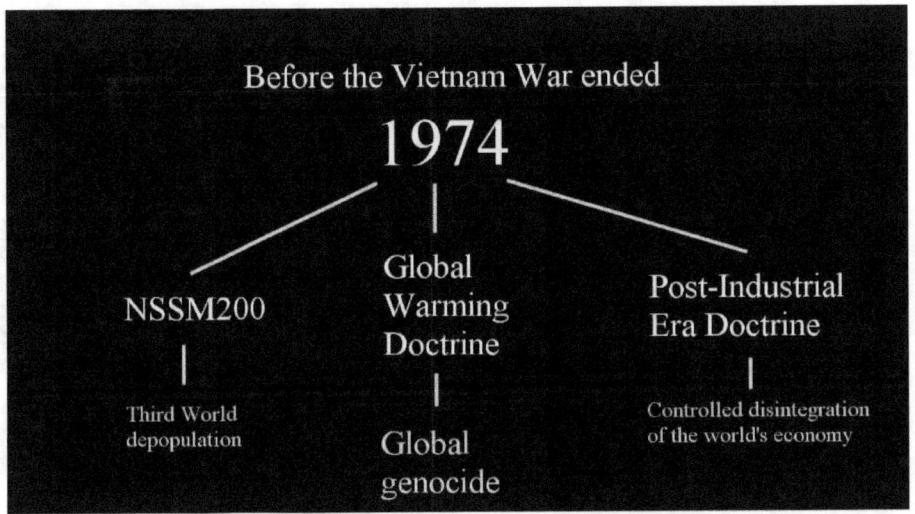

It may be interesting to note here that the 40-year Global Warming irregular warfare campaign was started up in 1974 out of the blue, and was organized from the halls of empire, in synchronism with other imperial projects of a similar nature.

After 40 years of psychological war

The Sleep of Reason Produces Monsters

from the Caprichos series - 1798
by Francisco Goya

The tragic fact is that the irregular warfare projects are all succeeding. Today, after 40 years of psychological war, much of the world sings the carbon scarecrow song from the heart, vigorously, and proudly - indicating that everyone is essentially asleep and dreaming of a world that has nothing to do with reality, but which is terribly scary. The dreamers will even fight you if you urge them to awake.

As if the illusions were real

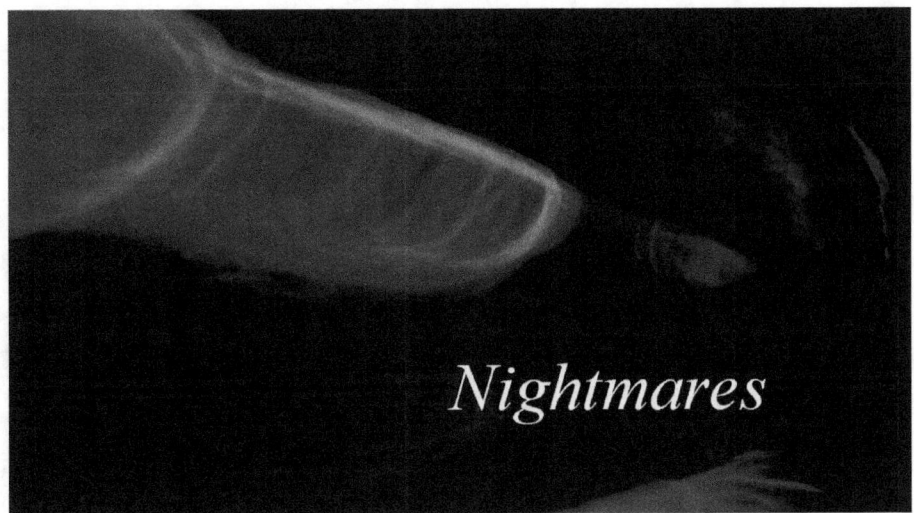

Nightmares

The dreamers have regressed to finding comfort in illusions, and comply with them as if the illusions were real. At this low-level state of dreaming, reality no longer matters.

The great global warming of the Earth

Hendrick Avercamp

For example, the fact that the great global warming of the Earth when the deep-freeze of the Little Ice Age ended, is not recognized as a cosmic phenomenon that is plain to see, even while it is massively supported as a cosmic phenomenon by real physical evidence.

Where real evidence isn't a factor

In the sleeping world, where real evidence isn't a factor, the re-warming of the Earth is attributed entirely to humanity's industrial action and its corresponding energy-rich living, which both produce atmospheric carbon gases, while it is hidden from the sleeping world that the life-critical carbon gases that have become emitted, are known to have essentially no effect on the Earth's climate whatsoever.

The cultivated blindness to reality

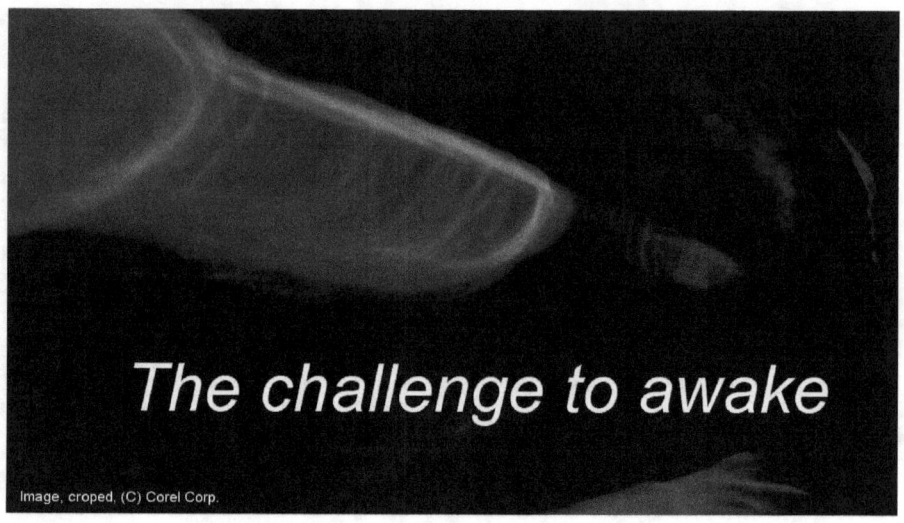

The challenge to awake

Image, croped, (C) Corel Corp.

The cultivated blindness to reality leaves one to wonder how deep the climate dreaming has actually become.

Shouldn't society rather open its eyes?

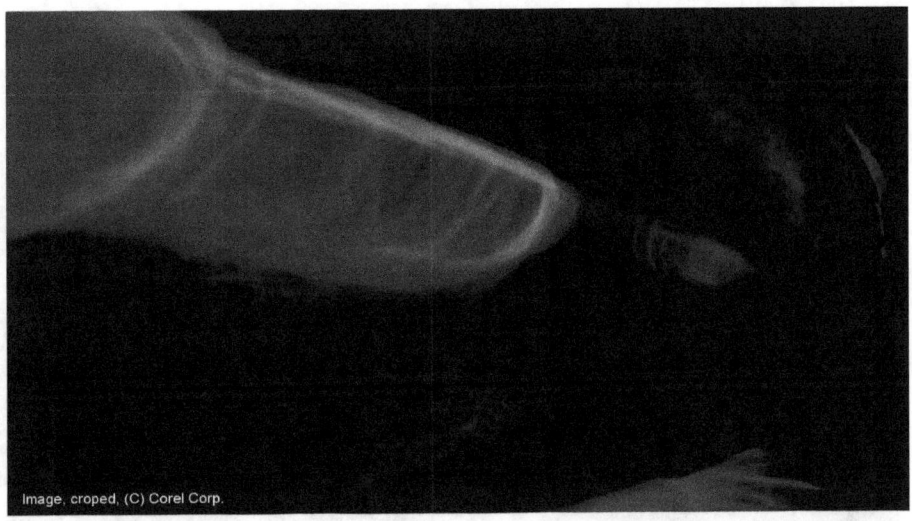

The obvious answer is, that what we see happening in the world is only possible when society is successfully inspired to close its eyes to the obvious facts around it, to commit itself to slumber and to ruminating over scare stories in a drama of nightmares about runaway global warming that overheats the Earth, melts all glaciers, floods the Earth with the ice turned to water, that raises the sea level, which nothing can stop than society committing suicide. That's scary indeed.

Shouldn't society rather open its eyes? When it would, a different world would come into view, than the one that it beholds in dreaming. But this awakening isn't happening, is it? 'It is prevented by the constantly ongoing psychological warfare that plays on what is not real.

#3 Climate Change Actions result from Solar Cosmic-Ray fluctuations.

World of reality, winged with honesty in science

The emerging scene that the awaking society would see, would be the world of reality, winged with honesty in science. It would become plainly apparent then that the massive global warming that has occurred in the period past the Little Ice Age, had absolutely a cosmic cause linked to the Sun standing behind it, rather than the Earth being super-sensitive to the magical CO_2 that is measured in parts per million and is vastly overshadowed by the greenhouse effect of water vapor that makes up, up to 97% of the greenhouse effect, which itself isn't the driving climate-change factor anyway, but is secondary in nature.

View of the Earth from ISS, Jan.4 2013, from over the mid-Pacific, 460 miles east of northern Honshu, Japan.

By viewing the world with open eyes and honest science, society would easily recognize that increased cloudiness reflects increased amounts of solar radiation back into space, which thereby becomes lost to the Earth, and vice versa. This means that all the big climate changes in the world are effects of ever-changing cloudiness, which in turn is affected by cosmic actions that are affecting the Sun, which is all physically measurable.

Electric ionizing enhances water-vapor nucleation

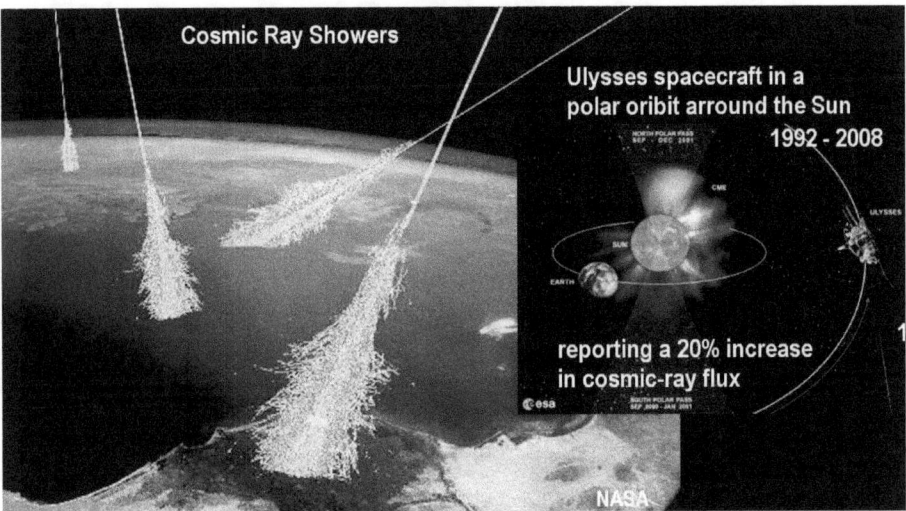

The evidence for the cosmic origin for climate action, is rather plain when one factors into the context the effect of cosmic rays on the cloud-forming process. All cosmic-ray flux has an ionizing effect in the atmosphere. The electric ionizing effect enhances water-vapor nucleation in the atmosphere that generates cloud droplets.

Solar cosmic-ray flux is massive

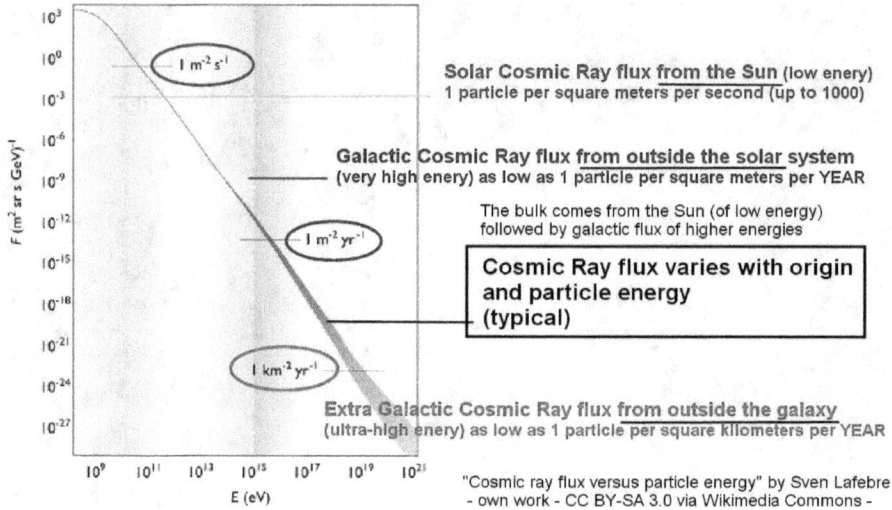

"Cosmic ray flux versus particle energy" by Sven Lafebre
- own work - CC BY-SA 3.0 via Wikimedia Commons -

It is measurable that most of the cosmic-ray flux that impacts the Earth, originates with the Sun in the form of solar cosmic-ray flux. The solar cosmic-ray flux is massive. It has been measured in the range of one to a thousand events per square meters per second, and it is constantly changing with cosmic factors that affect our Sun. Galactic cosmic-ray flux is rare in comparison, though more energetic. The Little Ice Age was demonstrably the direct result of larger volumes of solar cosmic-ray flux reaching the Earth at the time, which was a solar weak time that is evident in the lack of sunspots. The cause for the solar weak time, and the corresponding large solar cosmic-ray flux, is of course located far outside the solar system itself, where the sources are located that power our Sun, which also affect our Sun.

Cosmic rays are generated on the surface of the Sun

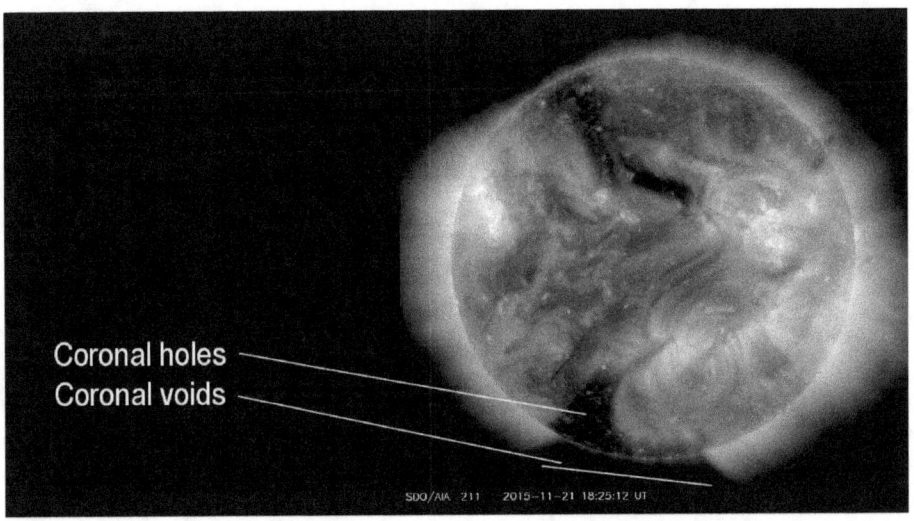

Coronal holes
Coronal voids

SDO/AIA 211   2015-11-21 18:25:12 UT

The changes in solar cosmic-ray flux can be dramatic. The flux increase typically results when voids develop in the solar corona. In these cases, the normally dense solar corona, which normally absorbs large amounts of the cosmic rays that are generated on the surface of the Sun, develops regional voids. When the corona weakens by weaker solar activity, the voids allow larger volumes of cosmic-ray flux to escape and reach the earth. How dramatic the voids can become, becomes apparent when the voids appear at the edge of the Sun. The coronal holes are thereby shown as huge gaps in the corona.

During the Little Ice Age

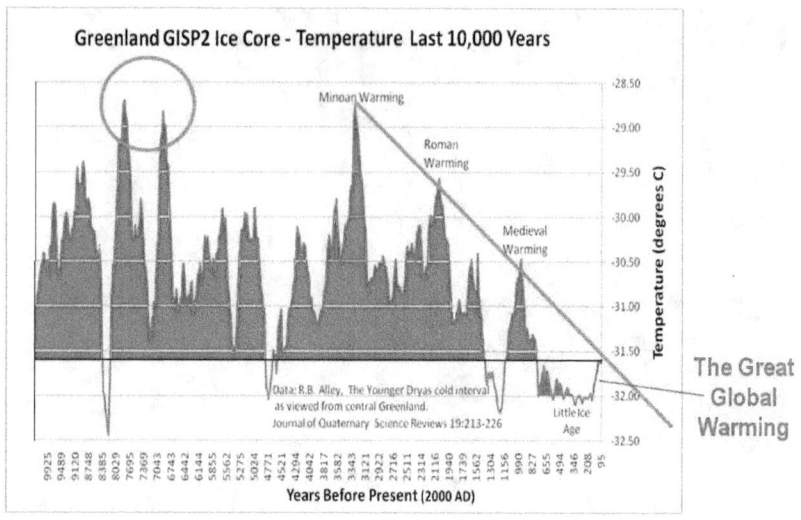

During the Little Ice Age, the Earth was so cold that rivers froze up and became skating rinks. Agriculture failed so dramatically that 10% of the population starved to death in some countries, up to 30% in the northern countries. The Sun had been in a state of prolonged, extremely-low solar activity at the time. The result was a 300-years period of Little Ice Age from roughly 1550 to 1850. For a large portion in this period no sunspots had been observed on the Sun at all, as if the Sun had gone to sleep, which, in a sense it had.

Sunspots during high-activity periods

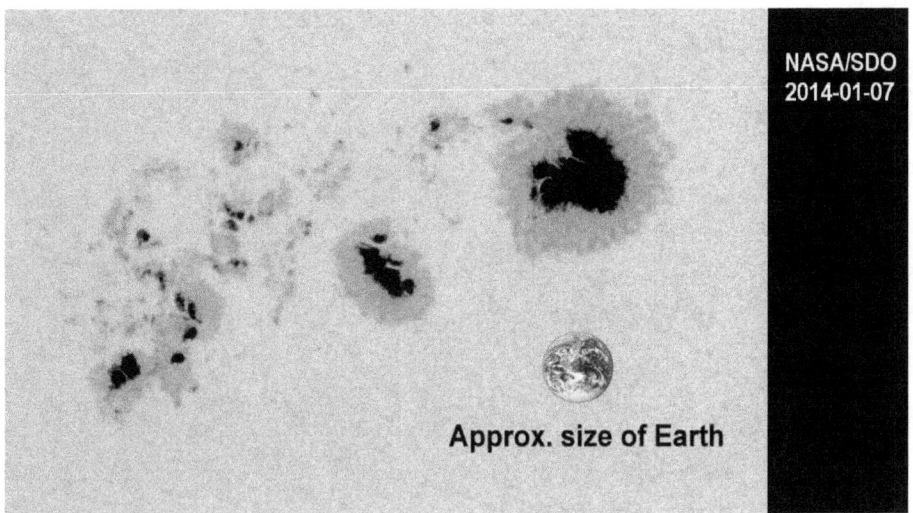

NASA/SDO
2014-01-07

Approx. size of Earth

Sunspots, typically occur only during high-activity periods when the Sun is intensely powered.

Sunspots are electric overload condition

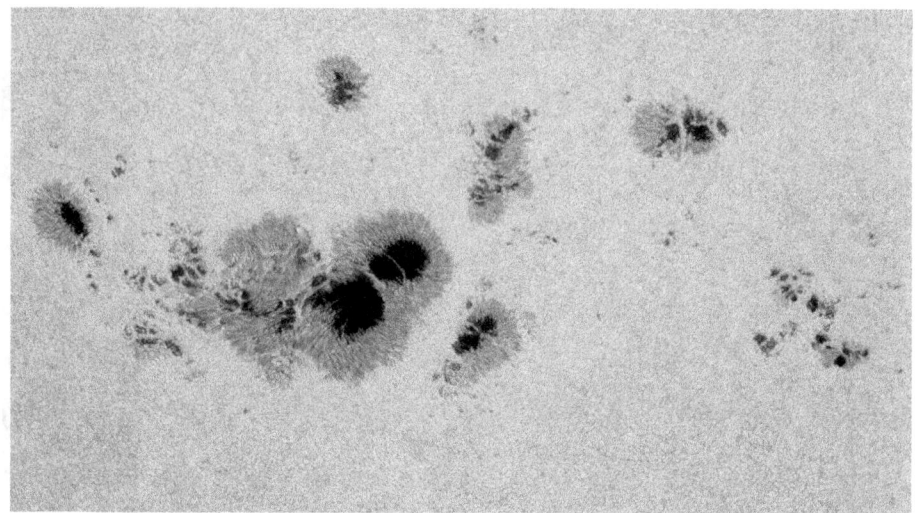

Sunspots are a type of electric overload condition that rips holes into the photosphere.

During the solar weak periods

The Sun in visible light
as seen through a dark glass

During the solar weak periods, the solar activity becomes too weak for sunspots to erupt. That's when fewer, or smaller, or no sunspots occur.

In weak periods, the Sun's corona is weaker

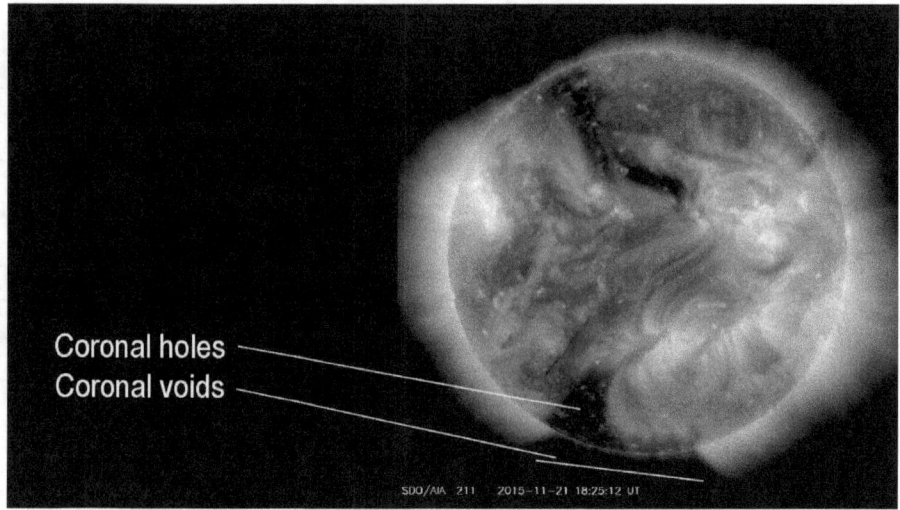

Coronal holes
Coronal voids

SDO/AIA 211   2015-11-21 18:25:12 UT

In weak periods, the Sun's corona is correspondingly weaker, so that holes develop more readily in the Sun's corona. The holes enable larger floods of solar cosmic-rays to penetrate the solar shield, which then impact the Earth.

Increased cosmic-ray flux increases cloudiness

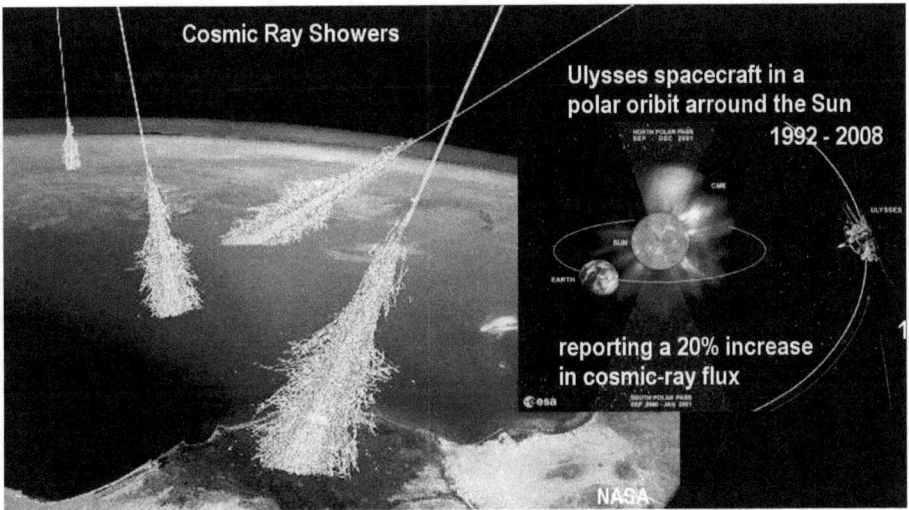

As I said before, the increased cosmic-ray flux hitting the Earth, increases cloudiness by cosmic-ray ionization of the atmospheric water vapor. This relationship has been verified with satellite measurements and laboratory experiments.

When artificial cosmic rays were injected

When artificial cosmic rays were injected into a test chamber, the resulting water vapor nucleation went straight up and off the chart.

A Little Ice Age results

View of the Earth from ISS, Jan.4 2013, from over the mid-Pacific, 460 miles east of northern Honshu, Japan.

And as I said before, increased cloudiness reflects a larger portion of the solar radiated energy back into space, which thereby becomes lost to us. The white top of the clouds are efficient reflectors of the radiated sunlight. When this happens for long periods, and happens intensively, a Little Ice Age results. That's what happened briefly in the 1300s, and more so between the 1500s and the 1700s into the 1800s.

The Sun was so weak

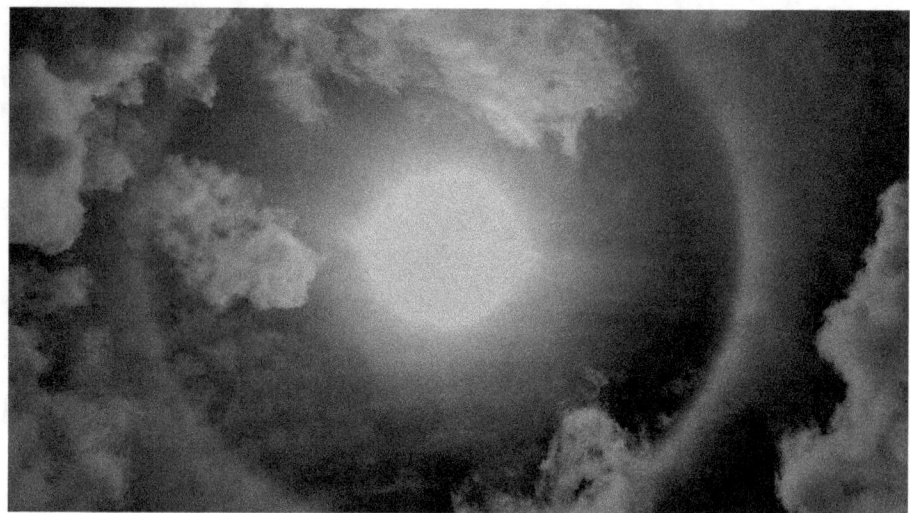

The Sun was so weak in these times - that just a few sunspots per year were visible in the 1600s when sunspots were counted and recorded. At the end of the 1600s only 50 sunspots had been counted in 28 years, in comparison with the 50,000 that are recorded in recent times for such a period.

The near total absence of sunspots

The Frozen Thames, 1677 - Abraham Hondius - wikipedia

The near total absence of sunspots stood clearly behind the coldest period in recent time, which became termed the little Ice Age that nobody wants to see happening again. The solar activity had been so low in this period that it had opened the door to high rates of solar cosmic-ray flux impacting the Earth with the effect of increased cloudiness and cold climates. It appears in retrospect that the Sun had barely recovered from its prolonged low-activity period.

What proof we have for a direct connection

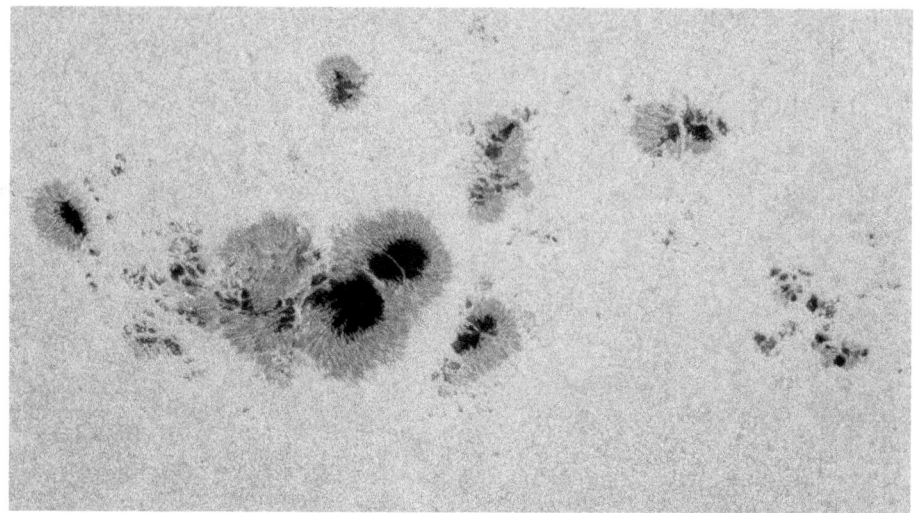

In this context it becomes imperative to explore what proof we have for a direct connection between the diminished sunspots and the colder climate on Earth. We know from historic observations that when the sunspots have disappeared, the climate became cold, and when the sunspots were back in big numbers the climate recovered.

Do we have physically measurable evidence?

Do we have physically measurable evidence for this?
Do we have hard evidence that the theorized link between the Sun and the climate on Earth, actually exists, and that this link really is the elusive solar cosmic-ray flux that no one can see? The answer is, Yes! The proof is provided by the Carbon-14 isotope.

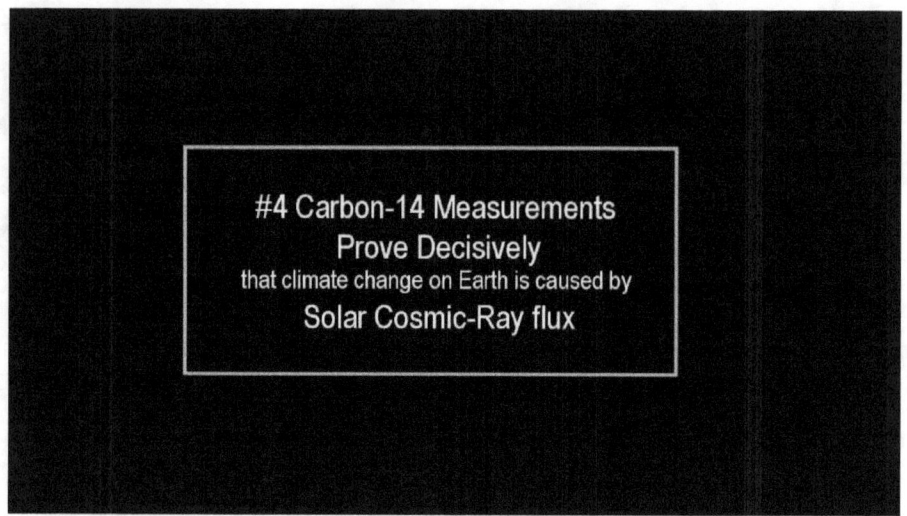

**#4 Carbon-14 measurements prove decisively that climate change on Earth is caused by solar cosmic-ray flux.

What is Carbon-14?

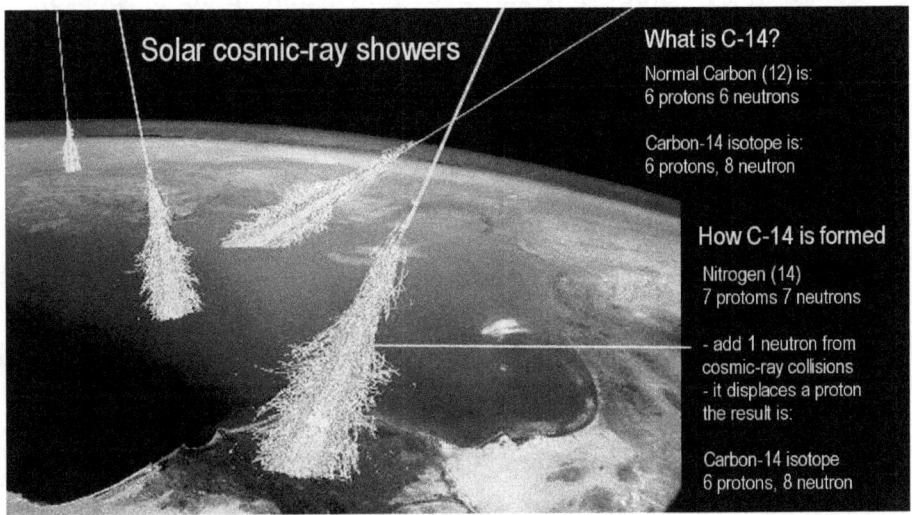

Solar cosmic-ray showers

What is C-14?

Normal Carbon (12) is:
6 protons 6 neutrons

Carbon-14 isotope is:
6 protons, 8 neutron

How C-14 is formed

Nitrogen (14)
7 protoms 7 neutrons

- add 1 neutron from
cosmic-ray collisions
- it displaces a proton
the result is:

Carbon-14 isotope
6 protons, 8 neutron

What is Carbon-14?

Carbon-14 is an unnatural carbon atom that is created by solar cosmic-ray flux acting on the Earth's atmosphere. The density of this unnatural isotope can be precisely measured. It thereby establishes a basis for measuring historic solar cosmic-ray flux. The natural atomic structure of the carbon atom is made up of 6 protons and 6 neutrons in its core, which renders it to be Carbon-12. In the 1940s the existence of a rare Carbon-14 isotope has been discovered, which is heavier. The Carbon-14 gets the designation, 14, because it has two extra neutrons attached. C-14 is rare in the atmosphere, in the order of two parts per trillion. But it is measurable.

The only known natural source for carbon-14 in the air, is the interaction of solar cosmic-ray flux with atoms in the atmosphere that by the collision emit a free neutron. the free neutron subsequently collides with an atom of atmospheric nitrogen. In this secondary collision, one of the protons of the nitrogen atom is displaced by the colliding neutron. In the process the nitrogen atom becomes transformed into a Carbon-14 isotope.

The Carbon-14 isotope is widely used for carbon dating purposes, because the unnatural precarious isotope decays with a half-life of 5,730 years. The rate of decay is used for dating organic objects. But this is not the only use for which the isotope is valuable. The Carbon-14 measurement is also used as a direct measurement for the solar cosmic-ray flux, which creates the isotope in the first place. The result is amazing.

Carbon-14 that can be measured

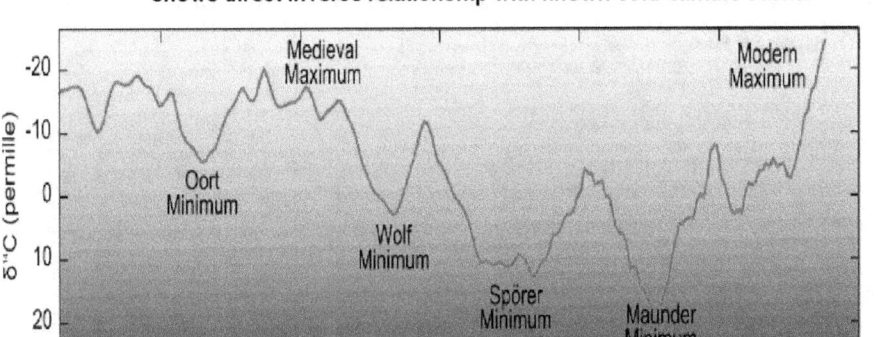

Changing <u>solar cosmic-ray flux,</u> measured in carbon-14 ratios, shows direct inverse relationship with known cold-climate events

When one plots the measured Carbon-14 values that one finds in historic deposits going back thousands of years, an amazingly accurate representation of historic solar-activity changes results. The chart shown here, clearly proves the inverse relationship between high solar cosmic-ray flux and low solar activity in sunspot counts, and of both coinciding with known periods of cold climate on Earth.

The fact that cosmic-ray interaction with the atmosphere converts nitrogen into Carbon-14 that can be measured, tells us unmistakably that the extremely cold period of the Little Ice Age that is coincident with the Maunder Minimum in sunspot numbers, was a period of extremely high cosmic-ray flux coming from the Sun.

Since it is known that increasing cloudiness, which increases with increasing cosmic-ray flux, directly increases the amount of incoming solar energy being radiated back into space, we have an unmistakable relationship established with the Carbon-14 measurements, that low solar activity increases solar cosmic-ray flux, which becomes reflected in colder climates, and vice versa. The bottom line is, that the Carbon-14 measurements present undeniable proof that all the huge climate changes that are known, have resulted from the cosmic changes that affect our Sun, and by nothing else.

Carbon-14 values line up with historic cold periods

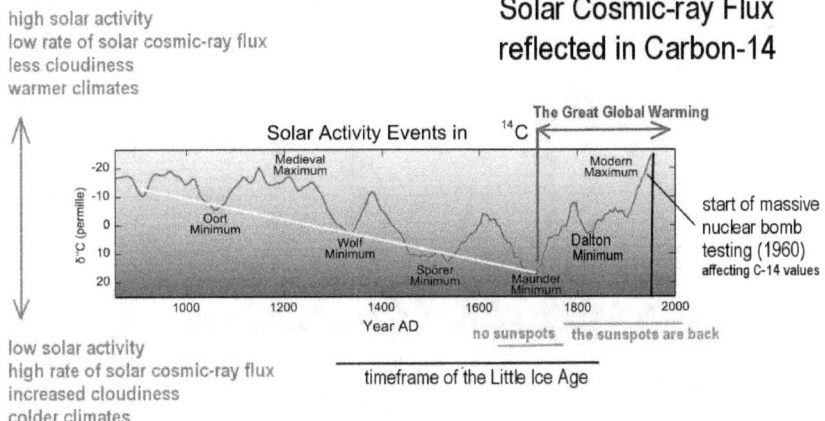

The measured Carbon-14 values line up so perfectly with the known historic cold periods, that the end result gives us on-the-ground measured comparative evidence that the long-term climate changes on Earth, both for the cold periods and the warm periods, are the direct result of the changing intensity of the operations on the Sun that reflect space-weather factors external to the solar system itself. Manmade CO_2 is demonstrated, thereby, not to be a climate factor. It also proves that the carbon-gas global warming hoopla employs a scientific fraud to vilify humanity.

It needs to be noted here that the solar reflection in Carbon-14 production was overshadowed from the 1960s on, by the effects of atomic bomb testing, after which the measurements became meaningless. It is reasonable to assume, however, that the established trend at the end of the measured period would continue, until other compelling cosmic factors would indicate otherwise, such as the diminishment of the solar wind NASA's Ulysses spacecraft has reported between 1998 and 2008.

Long doldrums of critically low solar activity

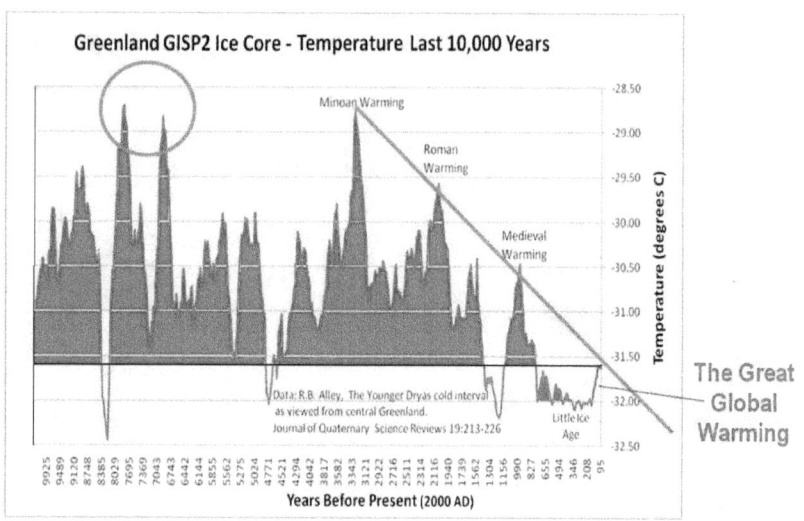

Greenland GISP2 Ice Core - Temperature Last 10,000 Years

Fortunately for us all, there had been enough plasma density remaining in the cosmic system at the time of the Little Ice Age, which may have remained in the background from the Medieval Warming period, so that the solar system was able to recover from its long doldrums of critically low solar activity. If the recovery hadn't occurred, very few people would be alive on the Earth today.

As it was, we were lucky. In the 1700s the sunspots began to re-appear. Cloudiness diminished. The Earth became warmer again. The Great Global Warming began.

Obviously, as I said before, the recovery of the Sun clearly wasn't caused by increased human activity, industrial processes, and carbon energy production, and so forth. Humanity has not yet achieved the capability to affect the Sun in any way whatsoever.

It would be wonderful if we had the capability

It would be wonderful if we had the capability, as human beings on the Earth, to invigorate the gigantic system of the Sun, because if we had this capability, it would enable us to avoid the next Ice Age that will likely begin in the near future, in the 2050s, according to a great volume of evidence at hand.

Affecting the Sun an impossible dream

Unfortunately, our affecting the Sun remains at the present time an impossible dream. This means that we have to face the coming Ice Age with open eyes, and face it as the universe is causing it, and that we react to the timing that the universe dictates.
The timing is critical here. The start of the next Ice Age is likely closer than we would wish it to be.
Let's take this realization back to the sunspots.
For 28 years in the middle of the Little Ice Age only 50 sunspots occurred, in comparison with the 50,000 that occur in our time in the same length of period.

It seems miraculous that the Sun recovered

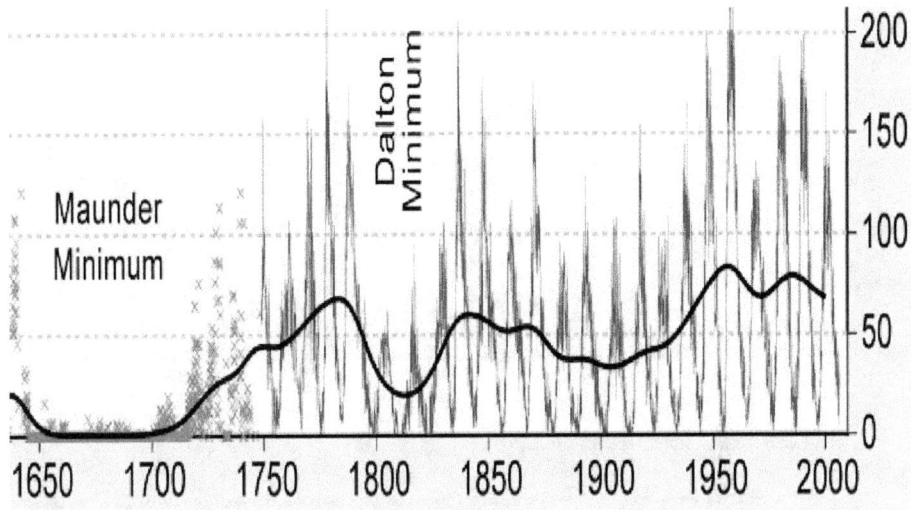

As I said before, it seems miraculous that the Sun recovered at all from this near-dead state. And even as it did, the recovery sputtered. It didn't get fully under way until after the Dalton Minimum had ended in 1850.

The minimal values in solar activity have been steadily diminishing

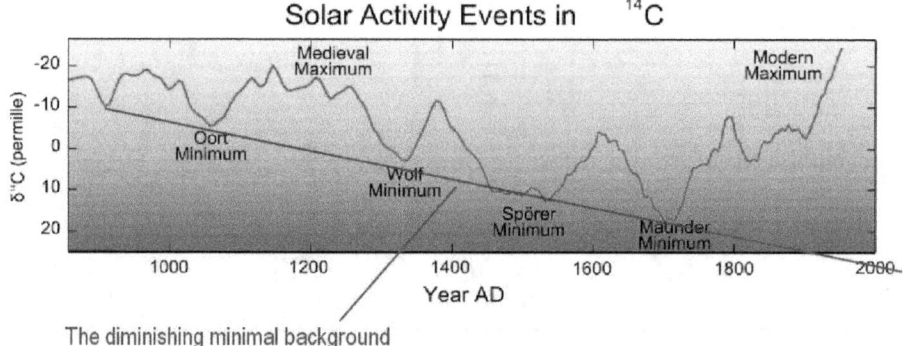

The diminishing minimal background

Just as in the long-term context the Carbon-14 data tells us that the solar cosmic-ray flux was extremely large at the time of the Maunder Minimum, which was the worst period of the Little Ice Age, the Carbon-14 data also tells us that the minimal values in solar activity have been steadily diminishing. It seems reasonable to project from the measured projection that the recovery from the next large solar minimal and its renewed Little Ice Age, may not happen at all.

We see the same down ramping

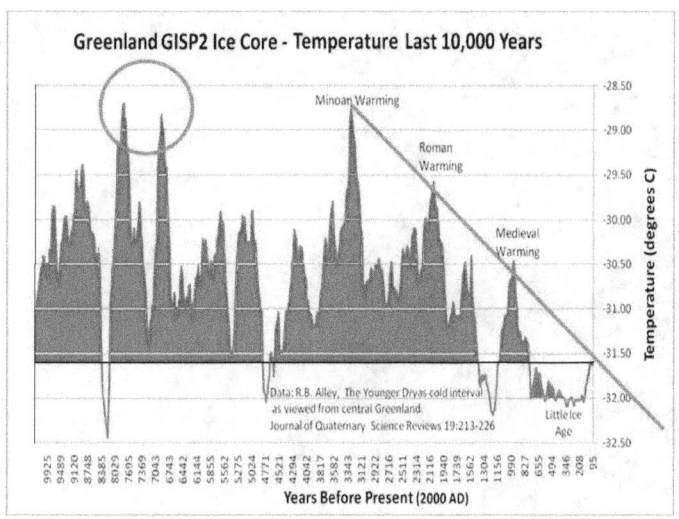

We see the same down ramping that we see between the minimal periods also reflected between the maximum periods. Both of these measured slopes project that a recovery from the next Little Ice Age is extremely unlikely, that instead a phase shift will happen that will take us into the inactive phase of the Sun where the surface plasma fusion stops and the next Ice Age begins.

Solar system may have been close to the phase shift

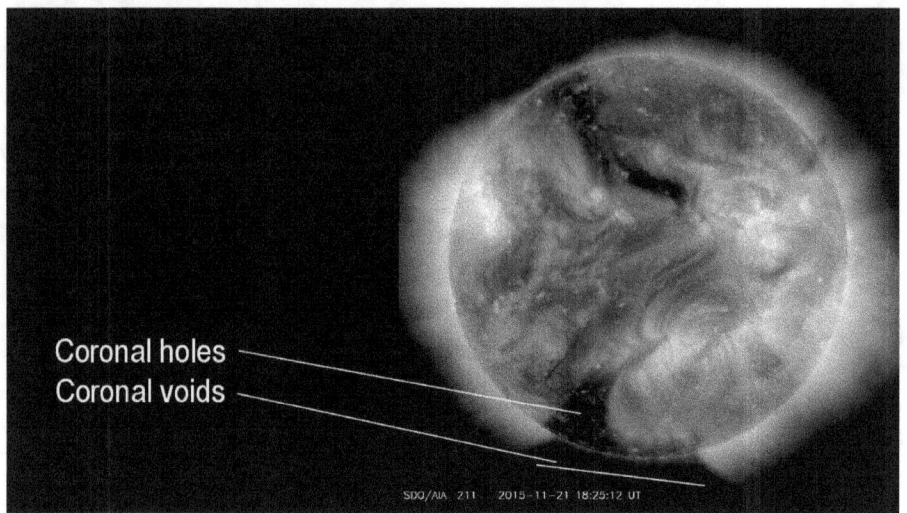

Coronal holes
Coronal voids

SDO/AIA 211   2015-11-21 18:25:12 UT

The high rate of solar cosmic-ray flux that we have measured for
the Maunder Minimum, tells us that the solar corona was not just
full of holes at this stage, but was extremely thin altogether. In
other words, the solar system may have been close to the phase
shift to the inactive state even then. We can expect the situation to
be much worse for the next Little Ice Age that is inevitably coming,
with which the next big Ice Age begins.

When the phase shift happens to an inactive Sun

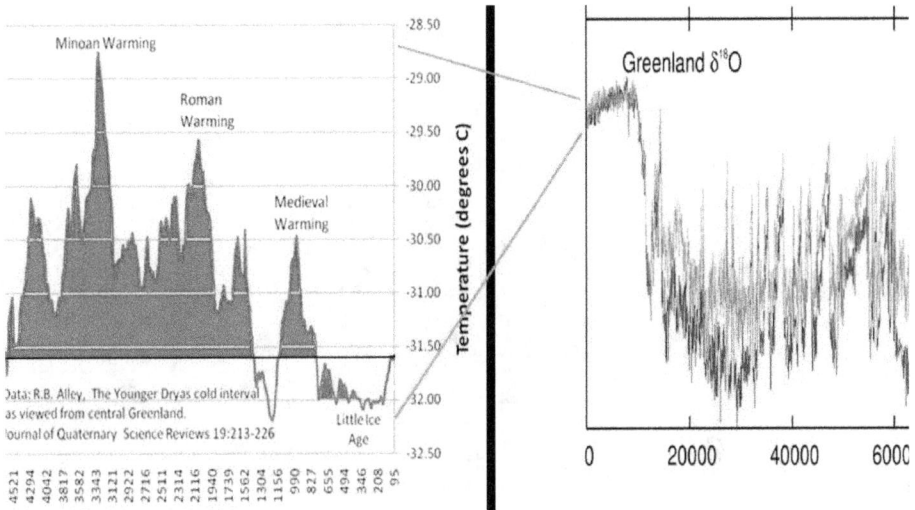

When the phase shift happens to an inactive Sun, and with it to the Big Ice Age, an entirely different platform for the climate on the Earth begins. The comfortable interglacial climate platform that we presently enjoy, will end, and the glacial platform will resume that we have never experienced in all remembered history, but which we have glacial records of. All of the climate experiences in the entire period of civilization will appear as nothing then and be overshadowed by the new reality of a cold climate that promises to be 40 times colder than the Little Ice Age had been, according to the ice core data from the pervious Ice Age.

That's what the down ramping points to

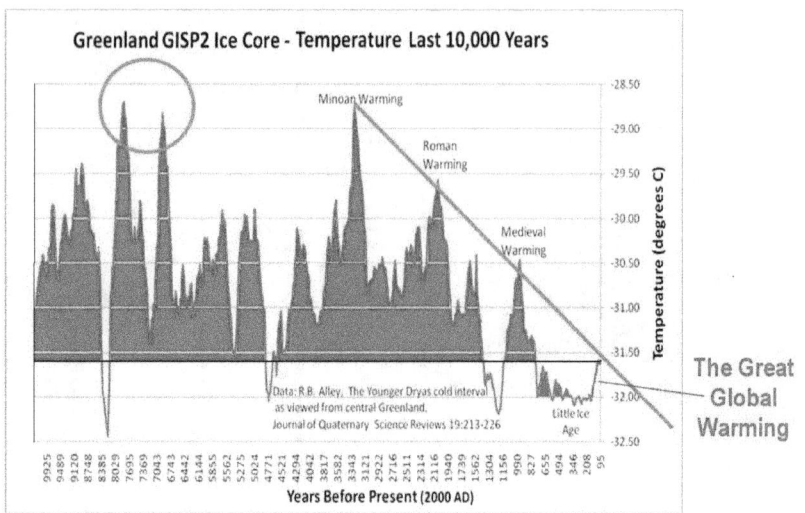

That's what we see as the result of the long-term down ramping.
That's what the down ramping points to as being inevitable.

Down ramping also in the small

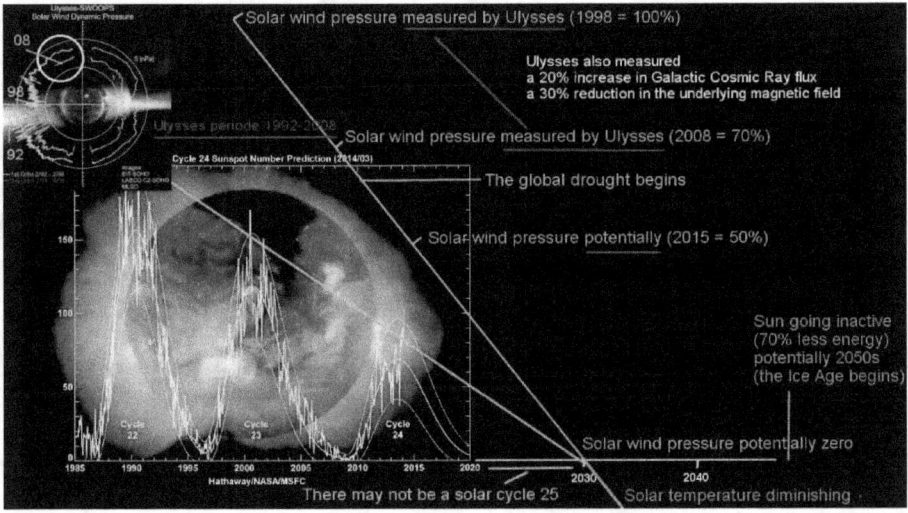

The down ramping is also already becoming evident in the small, in the short term. We see it reflected there as an ongoing down ramping of the individual sunspot cycles, and as a down ramping of the solar wind pressure that NASA Ulysses spacecraft had measured. Ulysses had measured a 30% drop in solar wind pressure in just ten years. That's huge. It tells us, that on the linear scale, the solar wind will diminish to zero in the 2030s, and that the Sun itself will diminish after that.

It is not unreasonable to recognize from this long diminishing of the electro-dynamic process, that the solar system will electrically collapse at some point on the diminishing slope, with the Sun going inactive, possibly in the 2050s, that then causes the next Ice Age to begin. We are extremely close to that. It may even happen sooner.

*The Next Ice Age is Near

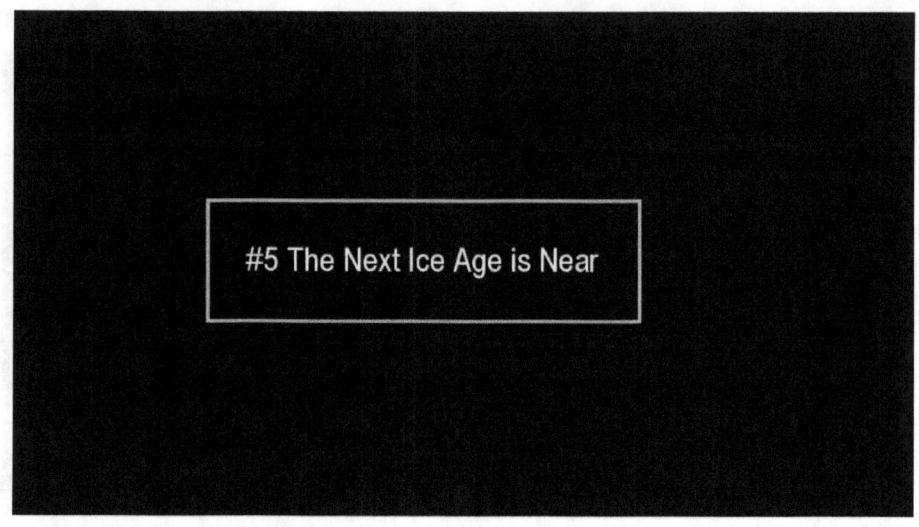

**#5 The Next Ice Age is Near

The liveable zone will shrink to a narrow band

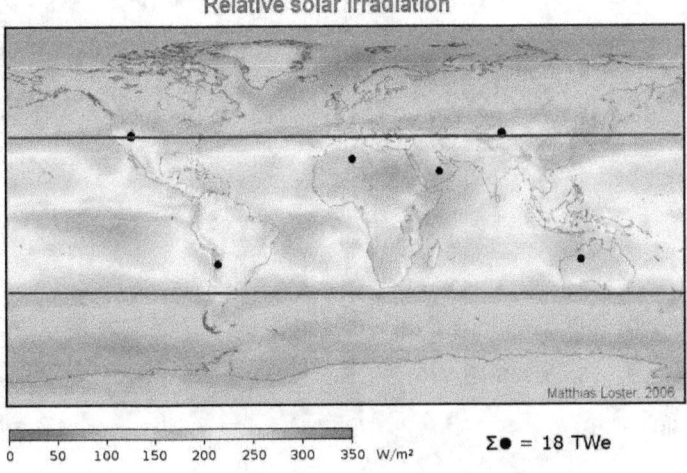

Relative solar irradiation

The climate on Earth will likely collapse somewhat less rapidly than the Sun itself will collapse, because large stores of thermal energy remain locked up in the oceans. Nevertheless, agriculture will collapse instantly in the diminished sunlight, except in the tropics where the solar exposure is two to three times greater. The northern regions, from Canada, to Europe, to Russia will most certainly become uninhabitable, perhaps equally rapidly, as the liveable zone will shrink to a narrow band centered on the equator. Whether humanity survives the transition, which may happen 30 years from the present, will depend on society's reaction in the present.

Thousands of new cities in the tropics

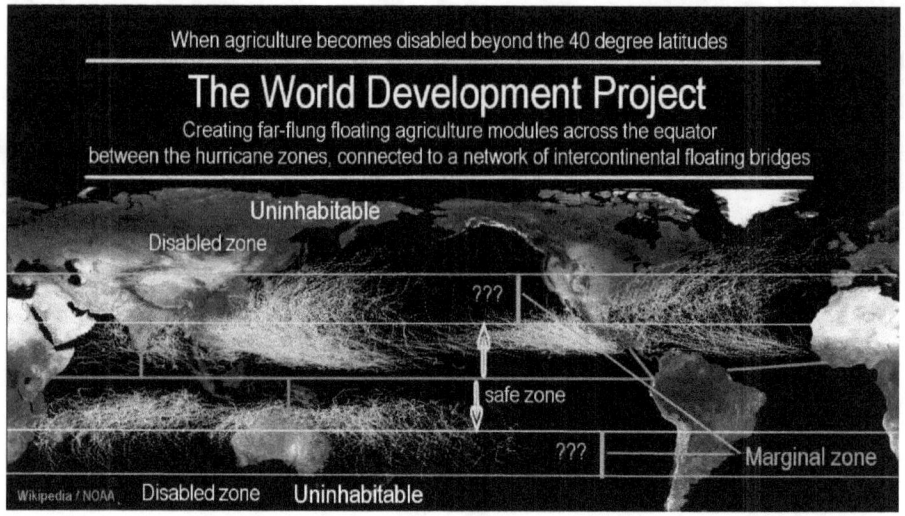

If the infrastructures are being built that enable us to relocate the northern countries into the tropics, with thousands of new cities being built in the tropics, and with new agriculture and industries being created there, humanity will have a chance to continue to live and prosper. If not, only a few million people will survive. The rest of humanity will die of the cold and by starvation. Will you and your children be among then? All children living today are so affected, and all adults who expect to live for more than 30 years. The question whether humanity will live or die will be settled by how humanity responds in the present to the Ice Age challenge, and the efforts that are made individually, and universally, to assure that the infrastructures for survival will be created.

So, it is up to you too, whether you and your children will have a future or die in agony. It is not my place to make this decision for you.

This is the stage where we are at today.

All but a few will die

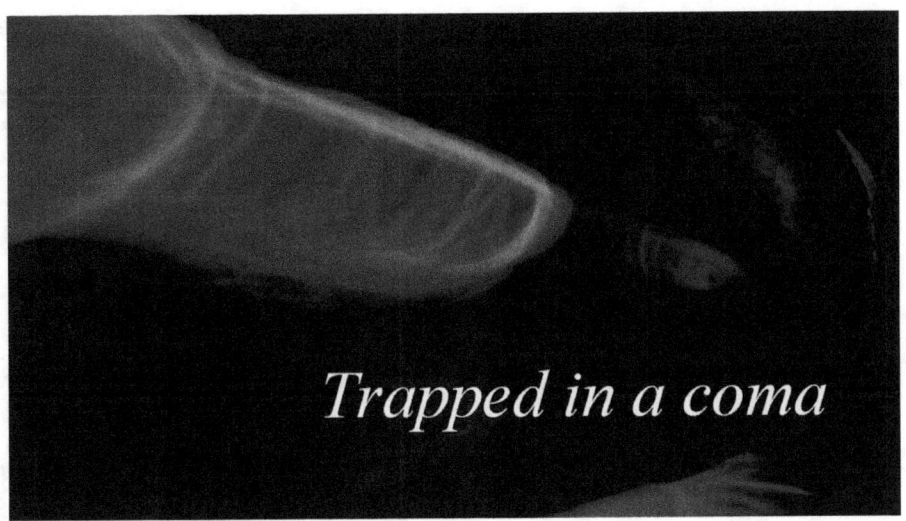

Trapped in a coma

Most likely, the decision by humanity to rescue itself will not be made, so that all but a few will die.
 Under the present regime of near worldwide irregular warfare against humanity - which has disabled humanity's soul, science, and culture on a wide scene already, and has demolished the very image of man as human beings, to the point that people become ashamed to be alive as a danger to the planet, both under the depopulation dogma and the carbon climate change doctrine - there is little power remaining in this state of coma in society, for society to wake itself up from its imposed coma and turn the ship around.

I am presenting a harsh assessment

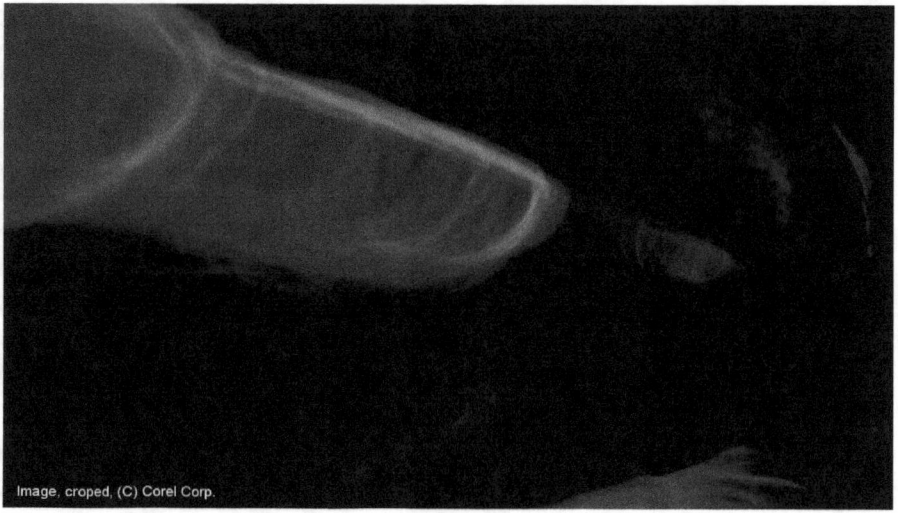

What I am presenting here may seem like a harsh assessment.
Unfortunately, the evidence supports this line of scientific
projection. Society is in a coma, asleep, dreaming, where nothing
much moves anymore. It lives the dreams it has been educated to
dream, such as the nightmare that manmade global warming is real,
dangerous, and is ultimately unstoppable, that it can only be
delayed, for which enormous sacrifices are demanded that become
an economic death sentence.

Infantile, akin to being ground into dust

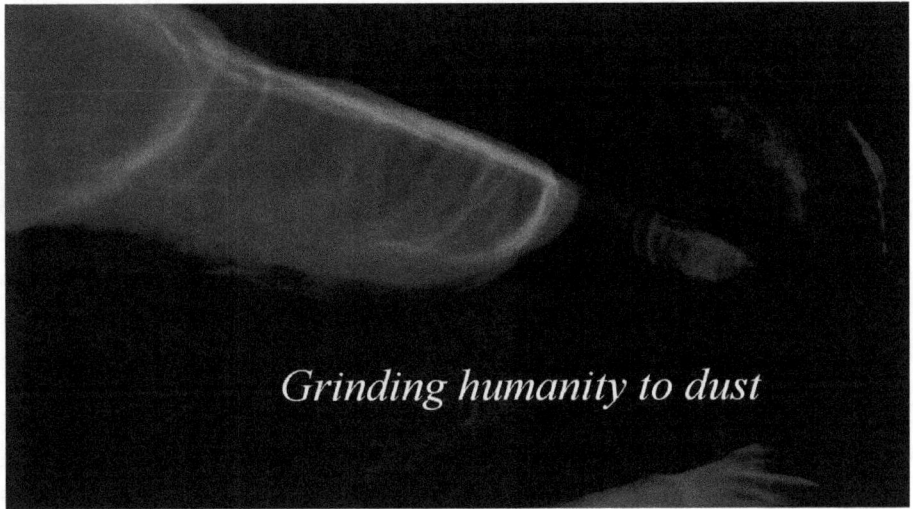

Grinding humanity to dust

The point is that once one is in a coma, nothing that is ruminated in the dreaming is grounded in anything real. The resulting dream-state level of thinking is at best extremely infantile, akin to being ground into dust.

Engineered sleep state has crippled science

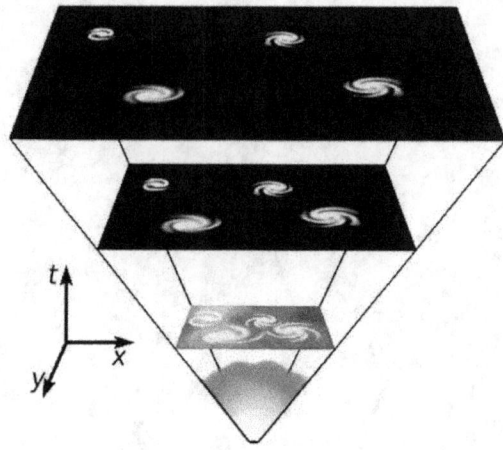

The engineered sleep state has crippled science. In the dream, reality has become foresworn. It is of no concern anymore.

As soon as the plasma galaxy model was proposed

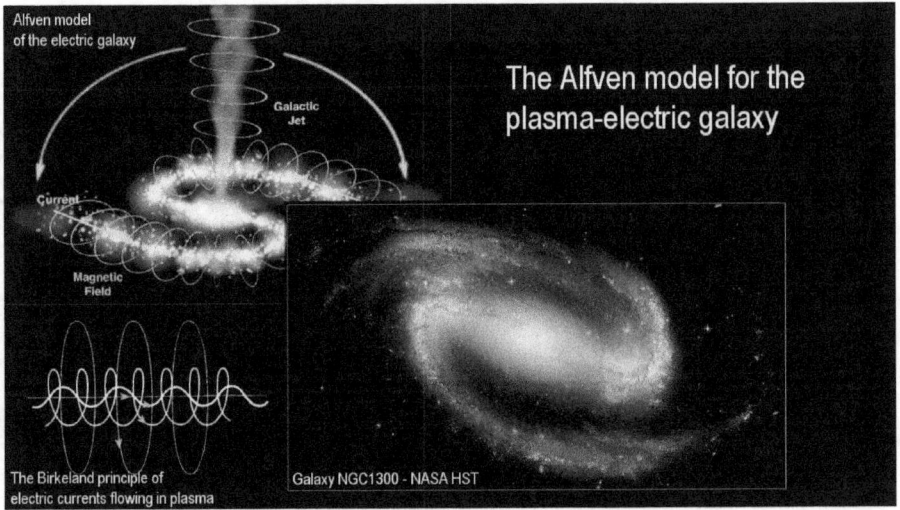

The Alfven model for the plasma-electric galaxy

Alfven model of the electric galaxy

Galactic Jet

Current

Magnetic Field

The Birkeland principle of electric currents flowing in plasma

Galaxy NGC1300 - NASA HST

For example, as soon as the pioneering concept of the plasma galaxy model was proposed, the Big Bang cosmology was hastily invented to bury the rational science concept with concepts where nothing is real.

The self-consuming hydrogen fusion Sun

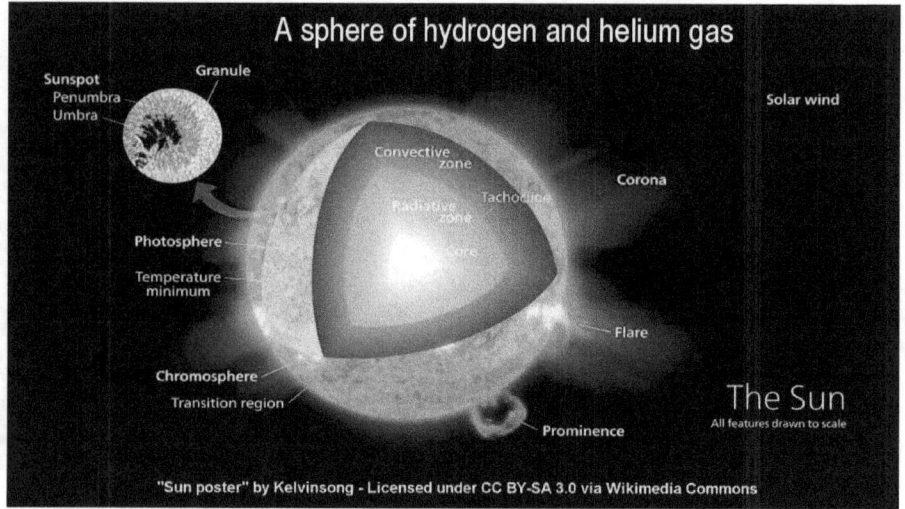

The model of the self-consuming hydrogen fusion Sun was similarly injected into the dream landscape where nothing is real, where what is deemed reality, is simply fabricated.

Academies of science don't believe in physics anymore

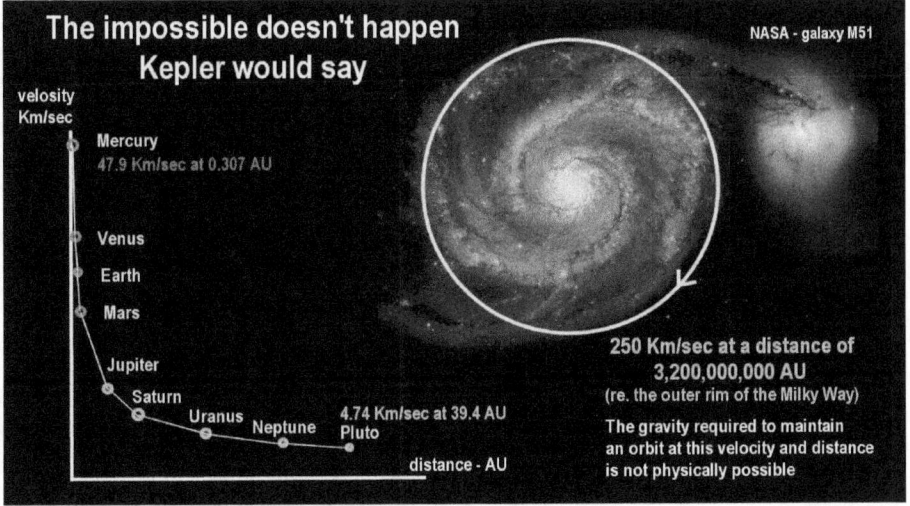

Another example of science inversion is the mythical galactic model of orbiting stars spinning around an imaginary galactic center. The resulting comatose state has been driven to such extremes that the academies of science don't believe in physics anymore.

A means to dishonour Johannes Kepler

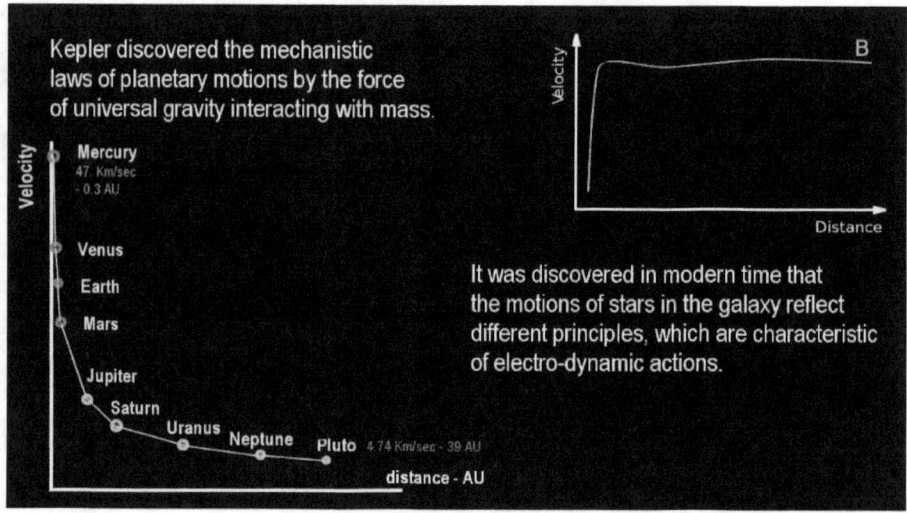

This sad diversion is evidently intentional, as a means to dishonour Johannes Kepler, one of the great pioneers in physical science, and to block the recognition of the Ice Age dynamics that is located in advanced science.

Truth dishonoured in dreams were nothing is real

Global Land–Ocean Temperature Index

The carbon-14 chart ends when atomic-bomb testing increased the C-14 values.

Carbon-14 index that represents solar activity changes

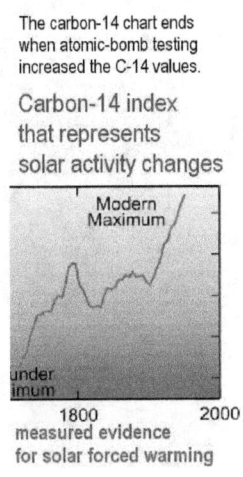

measured evidence for solar forced warming

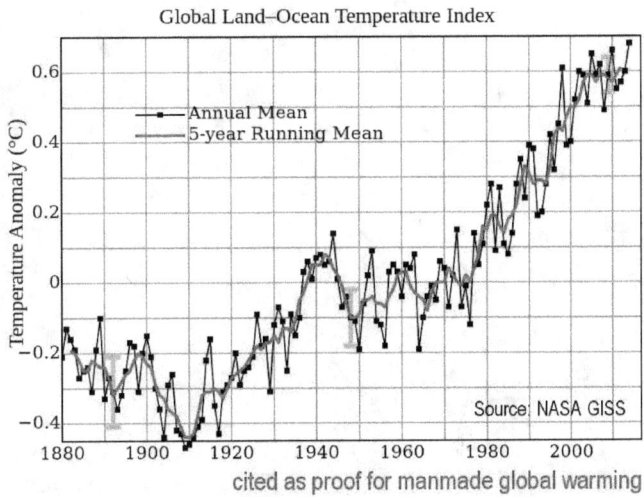

cited as proof for manmade global warming

Even truth itself is dishonoured in the dreams were nothing is real. The estimated global temperature increase on the right of the graphic is the scare crow of manmade climate change that has been paraded for decades to inspire humanity to destroy its energy production and scrap industrial development.

With the face of reality being kept hidden in modern irregular warfare, the simple fact is not being recognized that the increasing estimated temperature values, plotted by NASA and GISS, reflect nothing more than the known increase in solar activity that is reflected in the Carbon-14 measurements.

It needs to be noted here again that the solar reflection in the Carbon-14 values was overshadowed by the effects of the atomic bomb testing from the 1960s on. That's when the Carbon-14 graph stops. It is reasonable, however, to assume that the established trend has continued, probably until 1997, which the temperature chart indicates.

On the ground temperature measurements

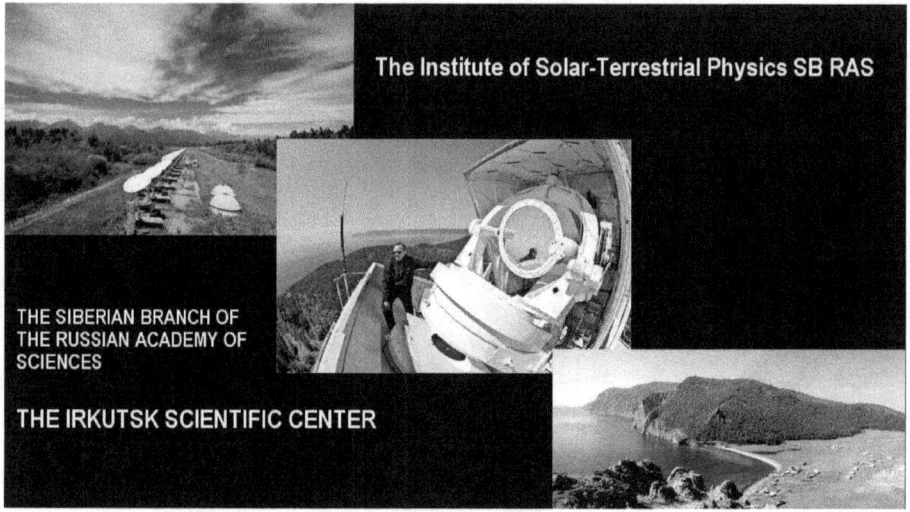

On the ground temperature measurements at the Institute for Solar-Terrestrial Physics in Irkutsk, indicated that a phase-shift to colder temperatures had begun in 1998, which is not reflected in the NASA graph shown here as the global reflection tends to lag behind the Irkutsk trend by three years.
The institute measured a drop in average annual ground temperature of 1.1 degrees in 1998, and an additional 0.5 degrees drop in 1999, and still a further 0.3 degree drop in 2000.

Measured cooling in Irkutsk, coincides

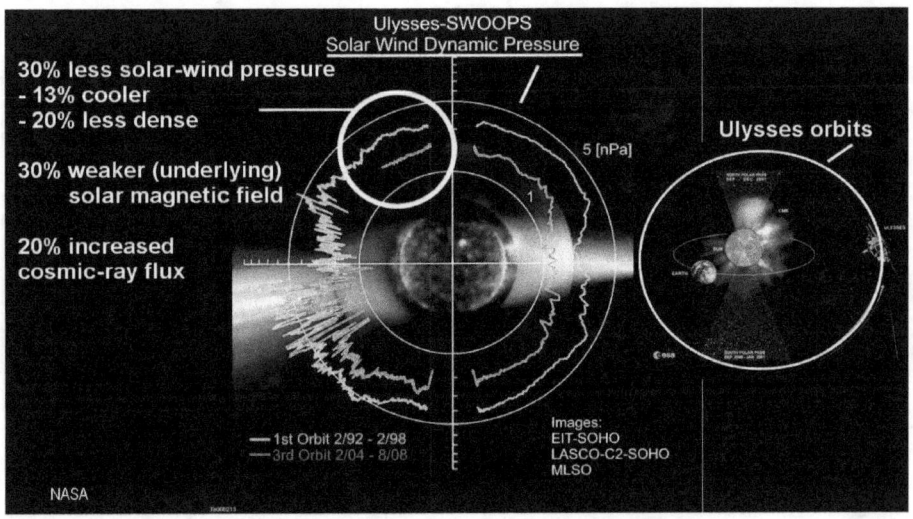

The on the ground measured cooling in Irkutsk, coincides in time with the 30% reduction in solar wind pressure that NASA's spacecraft Ulysses had measured in the space around the Sun on its third orbit beginning in 1998.

Carbon-14 exonerates humanity

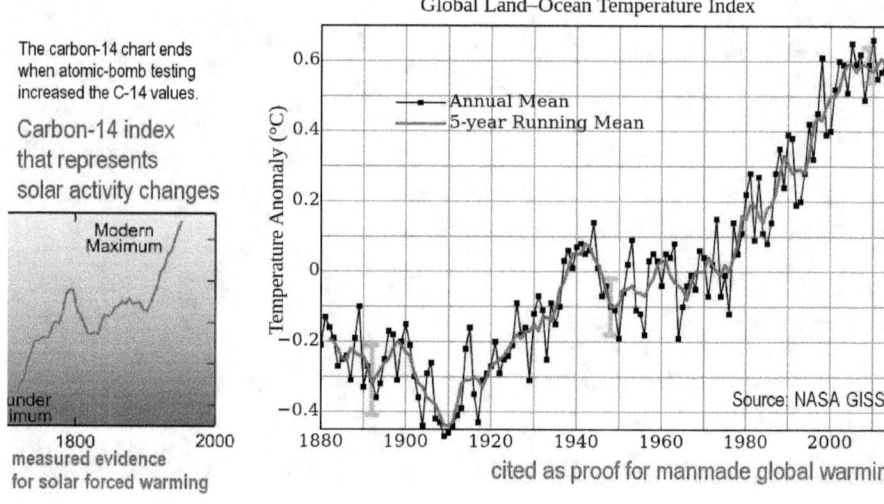

The carbon-14 chart ends when atomic-bomb testing increased the C-14 values.

Carbon-14 index that represents solar activity changes

Modern Maximum

under imum

1800   2000
measured evidence for solar forced warming

Global Land–Ocean Temperature Index

Temperature Anomaly (°C)

Annual Mean
5-year Running Mean

Source: NASA GISS

1880   1900   1920   1940   1960   1980   2000

cited as proof for manmade global warming

Both of these measurements, in conjunction with the Carbon-14 measurements, place the cause for climate change directly into the court of the Sun, and exonerates humanity from the charge of being a climate villain that the scare scenarios make it out to be.

As one would expect in times of war

COP15
COPENHAGEN
UN CLIMATE CHANGE CONFERENCE 2009

Lord Christopher Walter Monckton in Washington, D.C.
Head of the Policy Unit United Kingdom Independence Party
Photo © 2009 Joanne Nova (CC-BY-SA) Wikipedia

**Global Warming Fraud exposed by
Viscount Lord Christopher Monckton**

Of course, as one would expect in times of war, reality is never allowed to enter the theater of politics, as it would shut down the war. The real climate measurements are simply ignored in this theater, and replaced with fantasies. Not a word about anything real is allowed to be spoken in the theater of politics, which has become subsumed to the objectives in modern irregular warfare. For the same reason will the volumes of evidence for the coming Ice Age transition in the 2050s, remain hidden, including the underlying science of it.

Our prayers and highest hopes

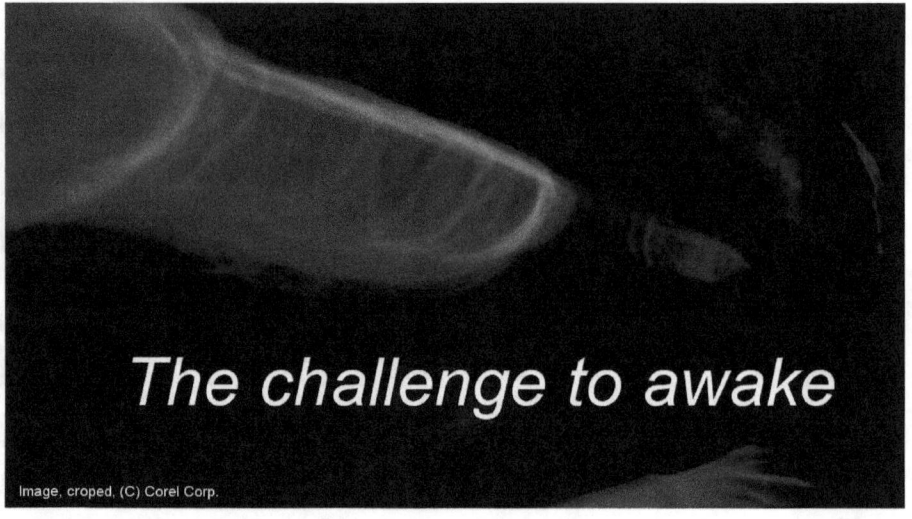

The challenge to awake

Image, croped. (C) Corel Corp.

In order to survive, society will have to dig itself out from its comatose state and rediscover what is real. This needs to include the rediscovering of what a human being is.
Our prayers and highest hopes should therefore be, that we may open our eyes to behold what is actually real.

The Earth's climate has forever been changing

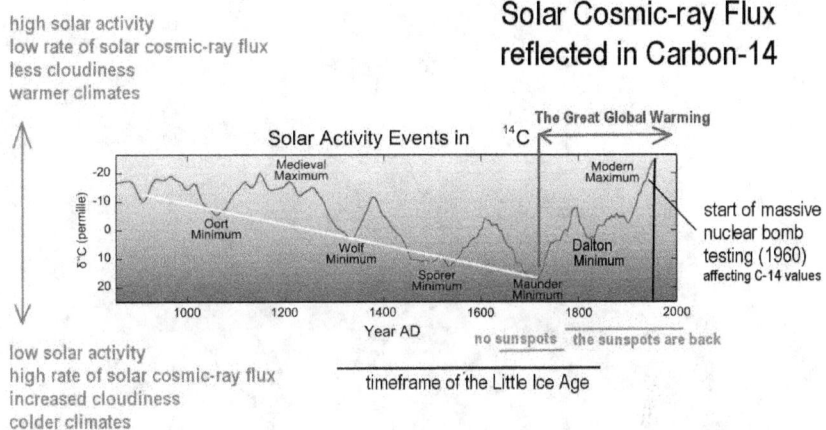

Then, when we get to this point and stop dreaming, and only then, will we acknowledge that while the Earth's climate has forever been changing, and will continue to change by the dynamics of its nature that is rooted in cosmic dynamics, the human presence and human action, which has not affected the Earth's climate in any way and never will, will be correctly understood. Then Man becomes exonerated on all counts from the charge of being a climate villain or a burden to the Earth.

The Milankovitch theory for the Ice Age cycles

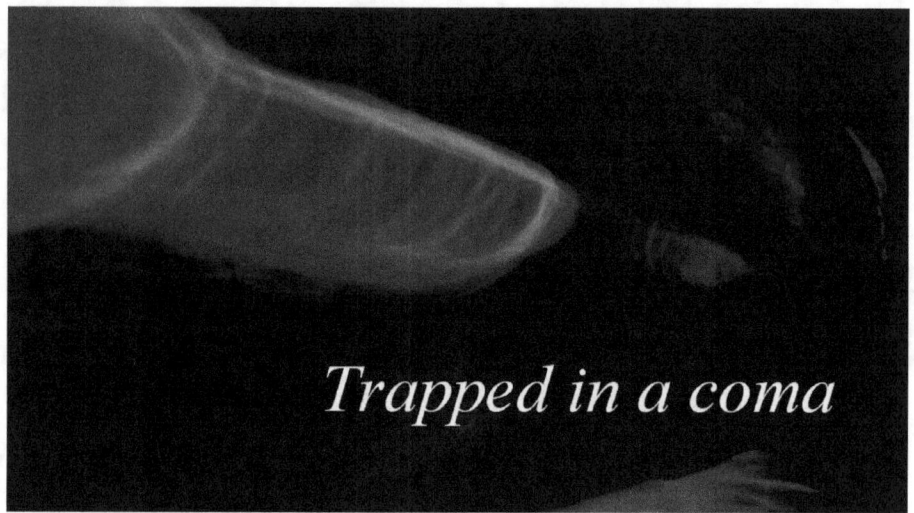

Trapped in a coma

The fact that much of humanity has been drawn into a state of coma, scientifically, where nothing is real, does not mean that all false science is intentionally false. Some science errors are the result of honest scientific mistakes, as for example the Milankovitch theory for the Ice Age cycles.

Other science mistakes are evidently intentional

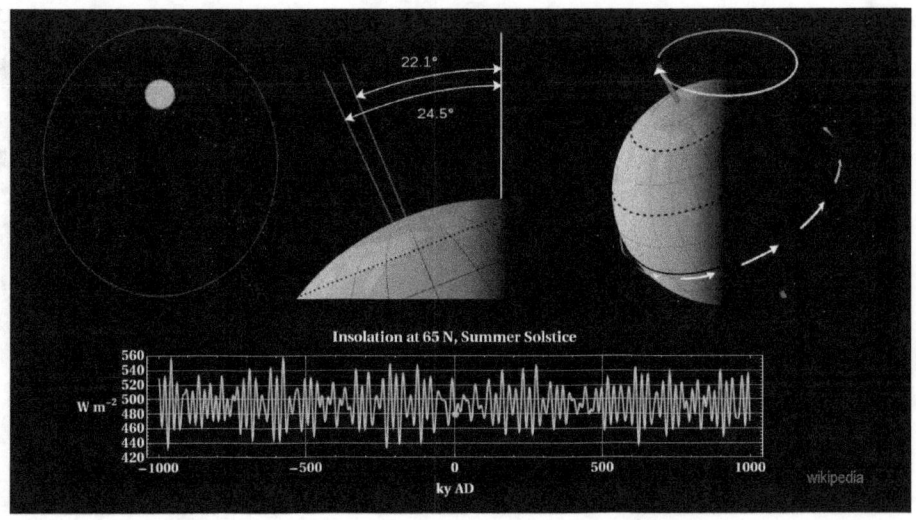

It had been long believed in earlier times, and still is believed by some, that the ice ages result from the combination of three types of minute long-term variations in the Earth's orbit around the Sun, with overlapping cycle times of 26,000, 41,000, and 100,000 years in duration. While these cycles cause minute variation in the hemispheric, and seasonal distribution of the solar radiation on Earth, the variations do not alter the total energy received from the Sun, whereby a global Ice Age can never really occur. In this case, a bit of common sense, and also Johannes Kepler's laws of planetary motion, invalidate the Milankovitch cycles theory, which is revealed thereby as simply a mistake.

Other science mistakes, however, are evidently intentional, as they defy known scientific facts.

Proof that CO2 Increase is NOT manmade.

**#6 Proof that CO2 Increase is NOT manmade.

It becomes scientific fraud to proclaim

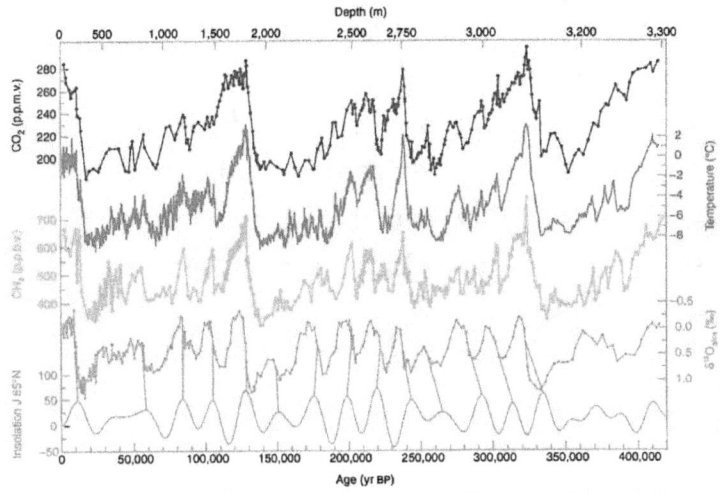

They typically defy clearly understood facts in order to drive the carbon climate fraud home, no matter what it takes. One of main elements of the carbon fraud is, that changes in CO2 cause climate changes. In this, ice core data is often cited as proof in support of the fraud. It has been discovered that Ice Age climates correspond with low CO2 levels, and the warm interglacial climates with high CO2 levels, and that the global warming that has occurred from the end of the Little Ice Age until 1998, has resulted from manmade additions to the global CO2.

This intentional science distortion of reality is plainly a form of science fraud, because it is an understood fact that cold climates reduce the global CO2 density, because in colder climates greater volumes of CO2 are absorbed in the cold waters of the oceans, while smaller volumes are evaporated in the tropical warm waters. The result simply means that changing temperatures drive the CO2 density changes. The lag time between the two is known to be in the range of hundreds of years, ranging from 300 years to roughly 800 years.

With this known, it becomes scientific fraud to proclaim that CO2 changes are driving the temperature changes, while the opposite is the case. The cause for the lag time of the CO2 response to temperature changes is understood.

CO2 changes responsive with a time delay

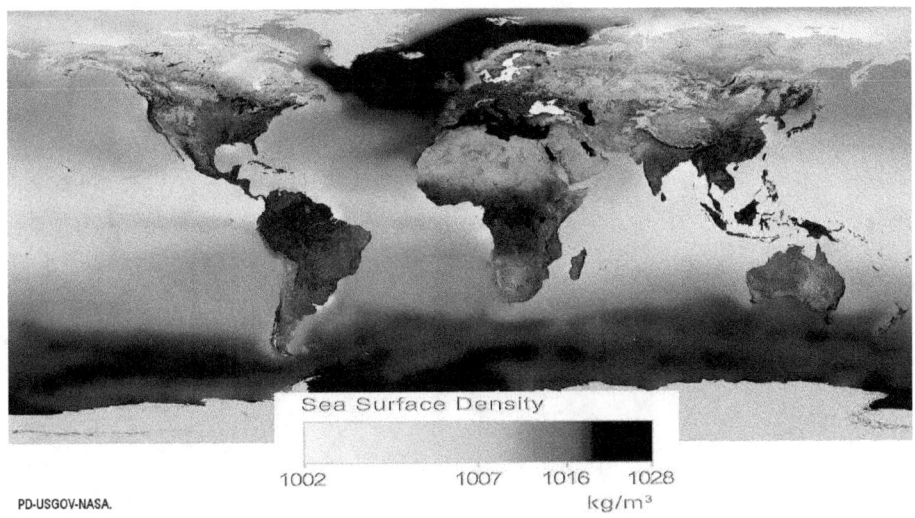

Sea Surface Density

1002   1007   1016   1028
kg/m³

The physical dynamics that render the CO2 changes responsive with a time delay, rather than being causative, are quite simple.

The fact that the CO2 in the air becomes primarily dissolved in the cold polar oceans, and cold polar oceans are having the highest surface density of water in the world, the cold CO2-rich water that are also dense polar waters, sink by their greater weight in the polar regions. They sink into deep pools from where they circulate across the planet in a system of interlocked ocean currents and carry their dissolved CO2 with them.

The cold polar ocean are CO2 rich, because cold seawater can contain slightly over 3000 parts per millions of CO2, close to 10 times the atmospheric density, which seems to explain why the polar oceans are extremely nutrient rich. CO2 is after all, critical for life, even for plankton in the oceans

The Antarctic deep current

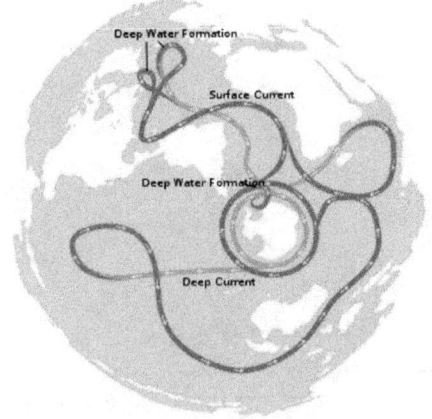

The ocean currents conveyor belt centered on the deep cold waters encircling Antarctica

"Conveyor belt" by Avsa - under CC BY-SA 3.0 via Commons -

The Arctic dense waters have an outflow from their pools that flows all the way to Antarctica where joins up with Antarctica's own deep pool that encircles the continent. The Antarctic deep current flows around the continent in the direction of the spin of the Earth, and slightly faster. As it does, two branches of dense waters spin off from it into two major streams.

The CO2 transit time in the range of 350 years

Thermohaline Circulation

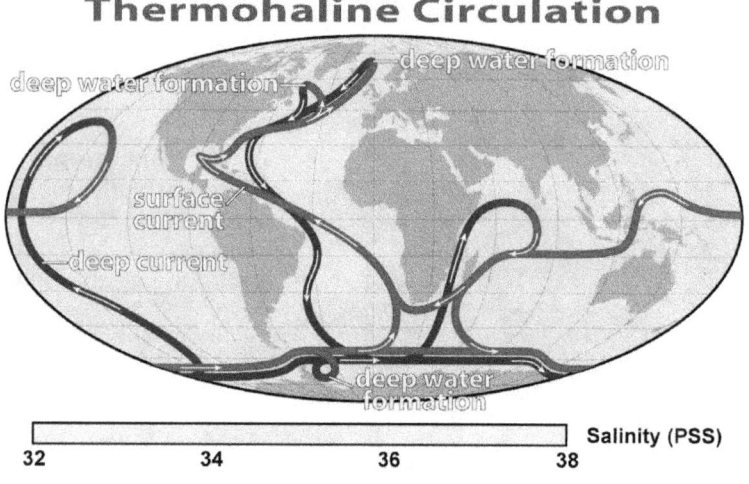

deep water formation

deep water formation

surface current

deep current

deep water formation

Salinity (PSS)

| 32 | 34 | 36 | 38 |

One of the branching deep streams, shown here in blue,
flows along the East Coast of Africa where the cold dense
waters warm up, rise to the surface, and in the process
relinquished their high concentration of dissolved CO2,
which become a part of the air again.

The second-deep branch flows into the Pacific where it's
CO2 is likewise relinquished. Because of the slow movement
of the currents it takes hundreds of years for the dissolved
CO2 to flow from the Polar regions back into the tropics to
become an atmospheric gas again.

For the deep current along the coast of Africa, the CO2
transit time appears to be in the range of 350 years, and for
the Pacific stream it may be in the range of 500 years. The
transit time from the Arctic to the Antarctic is believed to be
in the range of a thousand years. The long delayed CO2
response to temperature changes is evidently resulting from

the long transit times.

The atmospheric CO2 is not a static pool

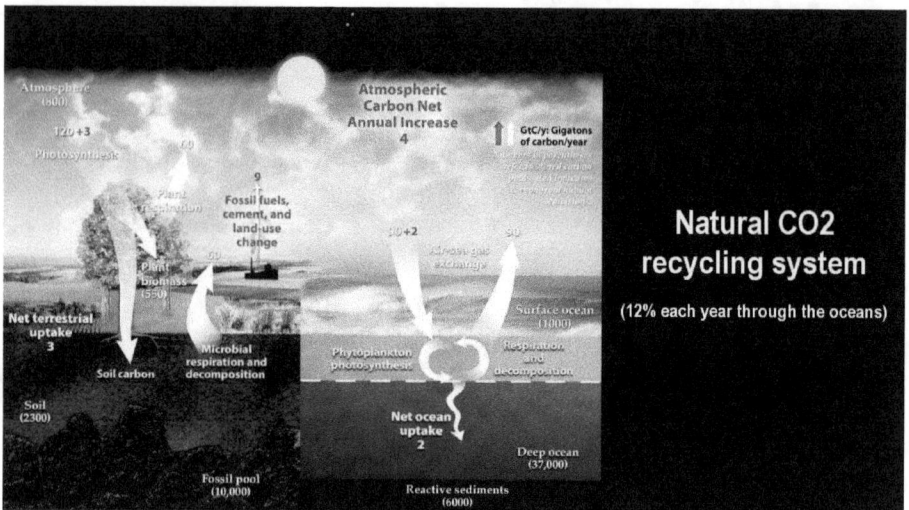

The atmospheric CO2 is not a static pool. It is constantly recycled. A large portion is recycled through the oceans at a rate of roughly 12% of the entire atmospheric CO2 volume, per year. It is known that the oceans hold 50 times as much CO2 than the atmosphere does. This means that 12% of the manmade contribution to atmospheric CO2 becomes absorbed into the oceans' recycling system, or essentially all of it within 9 years. The absorbed CO2 effectively disappears from the landscape for 350 to 500 years. It won't re-emerge until the end of its transit loop. For the CO2 from the Arctic waters, the transit time may be as long as a thousand years or more.

If every human activity that emits CO2

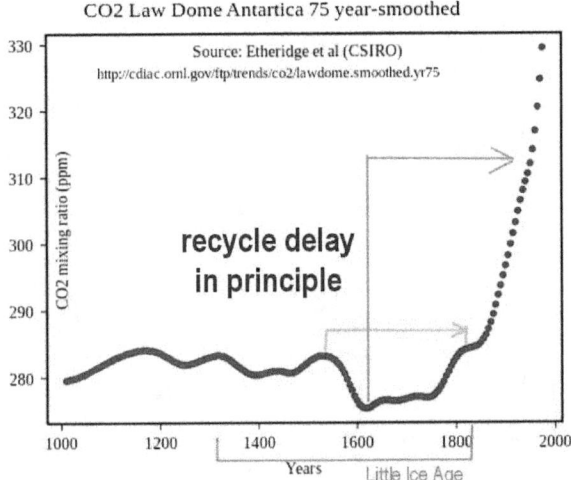

CO2 Law Dome Antartica 75 year-smoothed

This means that if every human activity that emits CO2 into the atmosphere was totally shut down, the increase of the atmospheric CO2 would continue, because a portion of today's atmospheric CO2 originated in the Little Ice Age and has been in transit through the recycling for 350 years.

Extremely large volumes of CO2 had been dissolved

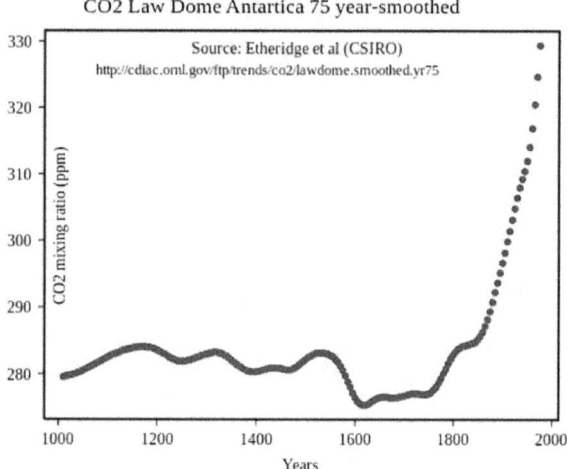

It is known that extremely large volumes of CO2 had been dissolved into the oceans during the colder times of the Little Ice Age, so much so that a major dip shows up in the ice core measurements of atmospheric CO2 of this time.

CO2 absorption in the Little Ice Age

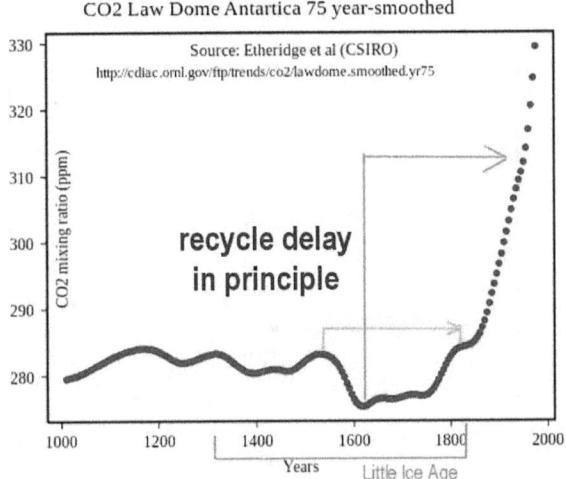

The CO2 increase that we see reflects the recycle time of the global conveyer belt system. The high rate of CO2 absorption in the Little Ice Age is brought back to us in the present. The Maunder Minimum is well within range for the recycle transit time to the coast of Africa. In the same manner is the CO2 from the more distant cold period of the Spoerer Minimum within range today via the outflow in the Mid Pacific.

Absorption dips from the last three big cold periods

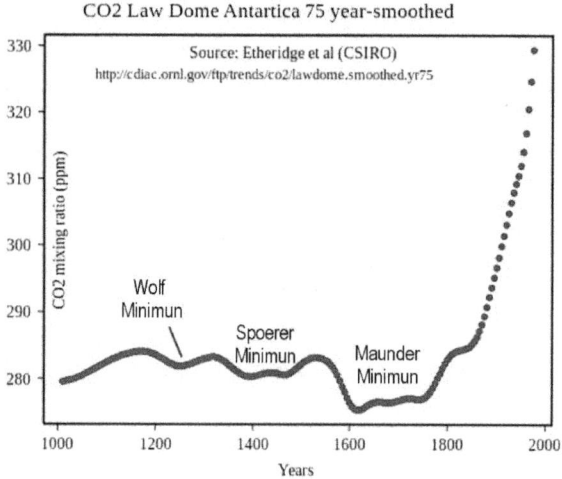

It would be surprising if the huge volume of CO2 that had been absorbed into the conveyor system during the big cold periods, would not reappear at the end of the transit cycle. This is precisely what we see happening, and what one would expect to see for an extremely slow recycling system. We see the absorption dips from the last three big cold periods in the ice core records in Antarctica shown here. Why wouldn't the absorbed huge volume of CO2 be returned to the atmosphere by the dynamics of this system?

110

The entire CO2 hockey-stick scare campaign

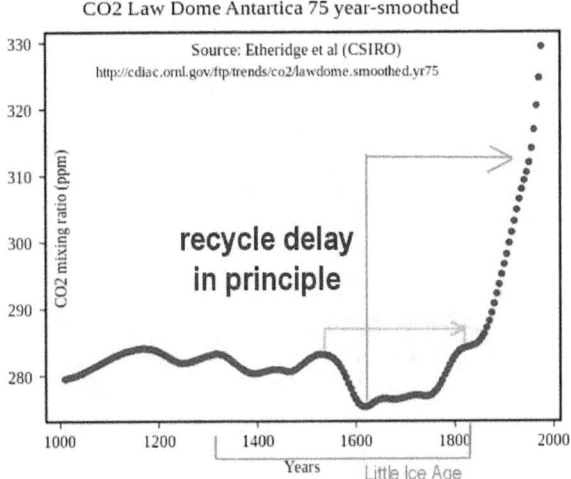

The answer is, that what we are seeing, in terms of the CO2 increase in modern time is clearly not manmade. This means that the result that we now see unfolding, is likely just the beginning of a much larger trend to come that would continue even if humanity ceased to exist. This also means that the entire CO2 hockey-stick scare campaign that vilifies humanity, is completely groundless, which adds another reason that renders the basis for the Carbon-Global-Warming terror project, groundless likewise.

The manmade-CO2 scarecrow card

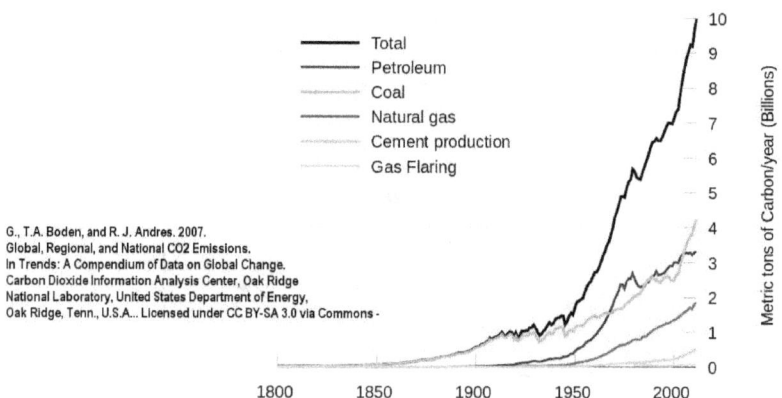

Sure, humanity is presently injecting roughly 8 to 10 billion tons of carbon gas into the atmosphere, which is in the range of roughly 1% of the total atmospheric CO2, the volume is insignificant as it all gets largely lost in the long-extended recycling process, rather than it hanging around, accumulating.

Considering that the manmade-CO2 scarecrow card with its typical hockey-stick rise in modern time is being blown up big for the vilification of humanity, while the major operating dynamics are all known that present opposite evidence, demonstrates that the CO2 scarecrow card is basically a fraud by intention that is deployed as a weapon in the games of modern irregular warfare against humanity.

The fraud vilifies humanity

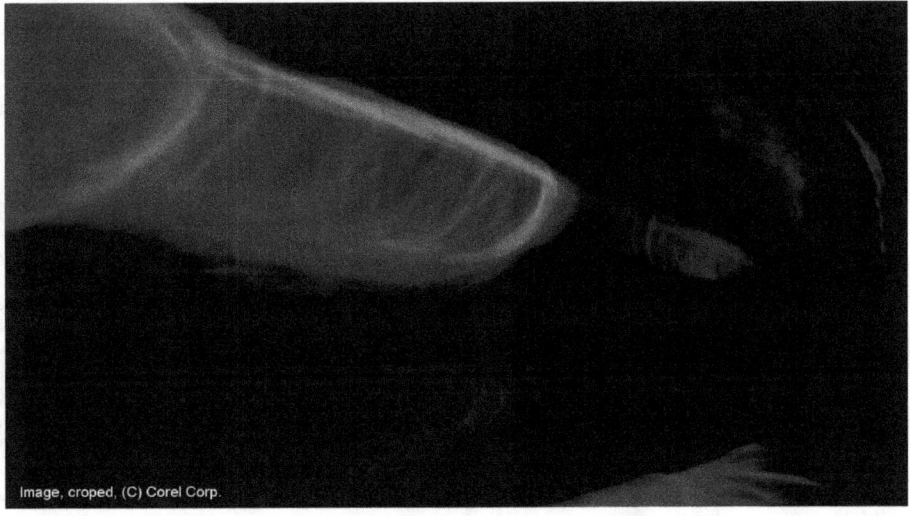

Image, croped, (C) Corel Corp.

The fraud vilifies humanity. It produces infantile responses in society, akin to a state of dreaming in which society can be easily controlled towards its self-destruction. This is the clearly demonstrated intention, which matches the intention of modern irregular warfare. The CO_2 fraud is then, evidently, just another card in the deck of terror of the modern irregular warfare process, where the physical reality is never an issue, but is typically hidden.

The 2015 terrorist acts in Paris

The 2015 terrorist acts in Paris, just weeks before the U.N. Climate Conference in this city, were likely perpetrated to assure that increased police-state measures would be implemented to prevent protest demonstrations and protest marches from occurring that might raise some points of truth against the climate scarecrow and awake a few of the comatose delegate assembled at the 120 million Euro extravaganza event that is designed to destroy humanity evermore deeply, with evermore strangling impositions, than humanity already is.

The modern irregular war against humanity has many faces and many cards in the deck, including the card of naked terror, while the war itself is singular, controlled and financed for a purpose, and the masters at the top are one. There is no such thing as spontaneous terror in the age of modern irregular warfare. If it was, it would be extremely rare. Terror is so radically foreign to human living and human identity, that it can exist only in organized form, cultivated, trained, financed, armed, and directed for a purpose. It exists as a card of a deck of many cards, which are all played in the way the strategists require.

Ultimately it will be irrelevant what happens in Paris, since the goal in modern irregular warfare is not centered on individual victories. The goal in irregular warfare is always the long-term objective. It is to grind the opponent, the opponent to empire, which is humanity, into the dust.

The biofuels scam is just a card in the deck

Mass Murder with Biofuels
a YouTube video

El Tres de Mayo, by Francisco de Goya - Wikipedia

The biofuels scam is just a card in the deck. In the name of reducing CO2 emissions from automobiles, high-value food products, such as corn and soy-beans, are converted into alcohol fuels to be burned. The project is a gigantic cultural warfare act from its inception. In the case of ethanol, the burning of it produces 7% less CO2 than octane, while the additional CO2 from the energy input for producing the ethanol fuel nearly doubles the total. Evidently the stated goal isn't the goal, but is a ruse that covers for the real goal, which is the destruction of humanity with the tool of mass murder.

Russell stated that wars don't murder enough people

Bertrand Arthur William Russell,
3rd Earl Russell, 1872 - 1970
1950 Nobel Laureate in Literature

Wikipedia

"Bad times you say, are exceptional, and can be dealt with by exceptional methods. This has been more or less true during the honeymoon period of industrialism, but it will not remain true unless the increase of population of the world is enormously diminished... War, so far, has had no very great effect on this increase, which continued through each of the world wars. [War] has been disappointing in this respect... but perhaps, biological war may prove more effective. If a Black Death could spread through the world once in every generation, survivors could procreate freely without making the world too full... The state of affairs may be somewhat unpleasant, but what of it? Really high-minded people are indifferent to happiness, especially other people's."

From *The Impact of Science Upon Society* (New York: Simon and Schuster, 1953) pp. 102-104

Lord Russell stated bluntly that wars are inefficient in that they don't murder enough people. He called for more efficient methods. The biofuels card plays in this direction. Biofuels are not an efficient new, net-energy resource, for the simple reason that the production of them requires nearly as much energy in total energy inputs than the fuel gives back. The biofuels card is only efficient in causing mass genocide in a hungry world with the mass-burning of food. The agricultural resources that are diverted from food production to biofuels production, would at the present stage, nourish upwards to 400 million people. In a world that has a billion people living in chronic starvation, the mass-burning of food leaves a trail of 100 million corpses in the wake every year by death from starvation. This horrendous death toll probably exceeds all the war-deaths of all the wars in history combined, repeated every single year. Maybe this is what Lord Russell meant, when he said that wars don't kill enough.

Committed to participate in the mass-murder holocaust

B100
bio diesel

Soybeans are considered by many agencies to be a source of complete protein

Large segments of the world are now committed to participate in the mass-murder holocaust by their own volition. They are dreaming with their eyes closed that this giant holocaust of murder - the largest in all of history - will save the planet from overheating.

Few people participating realize that they are the targeted victims

The face of folly

E85 = 85% ethanol fuel that is 39% less efficient

Very few people who are participating in this enormously wide genocide process, realize that they themselves are the targeted victims when they pull up to the gas pump and fill up with E10, or E15, or E85, or whatever the case may be, while the 100 million people who are dying of starvation, quietly in far-off lands each year, are merely a collateral expense for the creeping destruction of society as a whole, by grinding it into the dust, slowly and gradually, from within, as human beings, as an opponent to empire.

The principle of Fabius

The Roman politician and military general

Quintus Fabius Maximus

Founder of the principle of the Strategic Defence (280 BC - 203 BC) which became inverted into Modern Irregular Warfare

"The Shield of Rome"

But is this travesty really the method that had been pioneered by the Roman politician and general, Quintus Fabius Maximus, in defence of Rome against the vastly superior invading forces of Hannibal?

The principle of Fabius was, to simply disable the invading force, not to destroy it. Russia saved itself against Napoleon in 1812, by this principle. In modern irregular warfare this principle is turned upside down. Humanity is being attacked by it, instead of being defended by it. The resulting difference is as absolute as night and day. The ruling empire of today aims to destroy humanity to its very core, and boil it down to less than a billion to be kept like zoo animals, imprisoned and controlled. The problem is, for the empire system, that the inversion of a profound principle is recursive. Rome destroyed itself on this platform. It looted all the people round about till there was nothing left to loot, whereby Rome fell onto its own sword. The sword is recursive.

The current world empire without a name

The current world empire that operates without a name, or has many names, is already dying by its recursive sword. Its financial system that has looted the world, is functionally already bankrupt. The naked-terror card is played evermore furious now, on this ground, which by it being played, hastens the masters' own demise. No force in the world can prevent the fall of their system. However, humanity does not need to fall with them.

Tickets for Survival

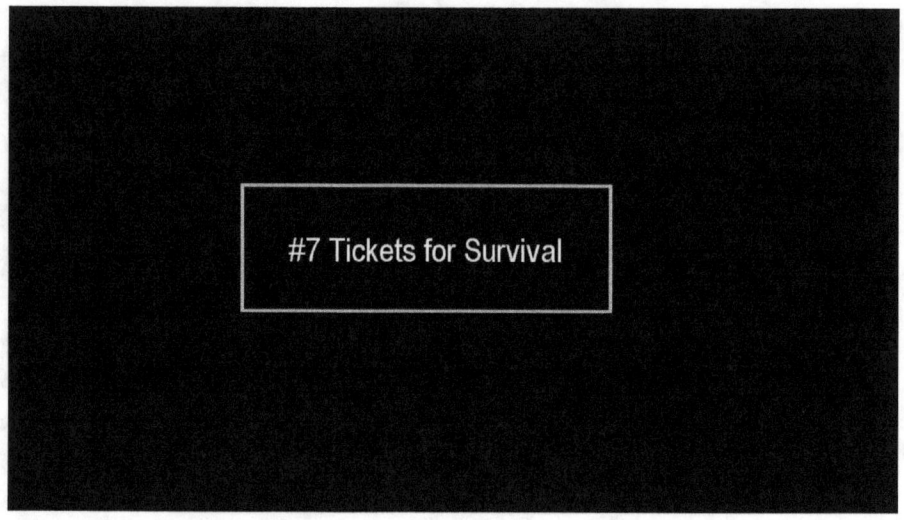

**#7 Tickets for Survival

The Fabius principle of the strategic defence

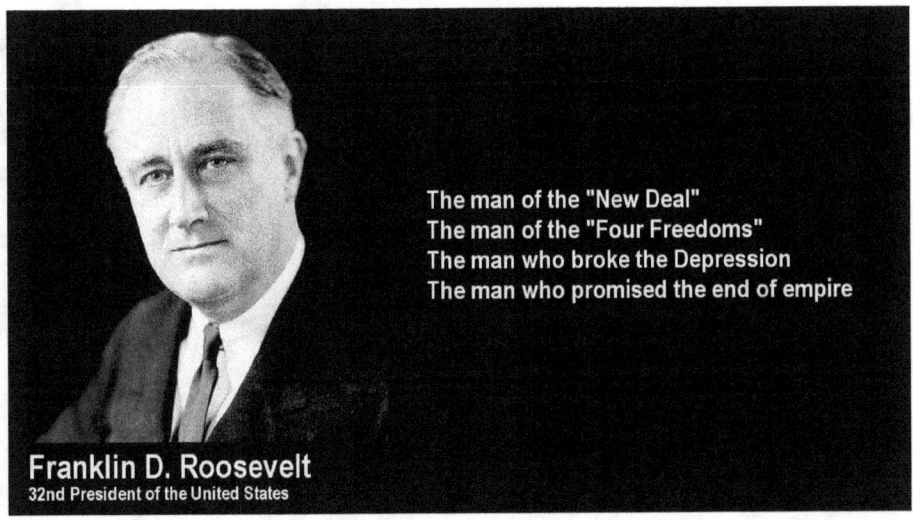

The man of the "New Deal"
The man of the "Four Freedoms"
The man who broke the Depression
The man who promised the end of empire

Franklin D. Roosevelt
32nd President of the United States

The Fabius principle of the strategic defence can be successfully applied by humanity in its defence against empire. The principle virtually guarantees its success. A move towards it is already on in the USA, to apply the Fabius strategic defence principle to save the nation. The strategic defence principle had previously been applied in the form of the Glass Steagall bank-system protection legislation, in 1933, by President Franklin Roosevelt. In recent years a movement has begun in the USA to restore the Glass Steagall legislation for the strategic defence of the nation. In doing this America writes itself a ticket for its survival.

Strategic defence against the Carbon Global Warming card

Wheat harvest on the Palouse, Idaho, USA

Food for eating!

wikipedia

A similar strategic defence is possible to be achieved by society against the Carbon Global Warming scarecrow card, the Biofuels scarecrow card, and the Depopulation scarecrow card. This can be done by simply looking at the reality, which is, that there is nothing standing behind these scare cards that isn't a pure fabrication on every point as this video has demonstrated. On this path, humanity as a whole, writes itself a ticket to have a future once again, which currently is not even a concept anymore.

The strategic defence that Fabius Maximus had pioneered

Annihilation is assured

500,000 times
Hiroshima
in one hour

Castle Bravo - the first U.S. test of a dry fuel thermonuclear hydrogen bomb - March 1, 1954 at Bikini Atoll, Marshall Islands

The principle of the strategic defence that Quintus Fabius Maximus had pioneered might ultimately be the only workable platform in humanity's defence against nuclear war. Nuclear war is designed from the ground up to function as a scarecrow card in the deck of modern irregular warfare. The scarecrow cards are each one designed in its individual way, to keep sanity away from the fields of the theatres of empire, and to render humanity stupid and asleep. The principle of the strategic defence must therefore cover the entire deck of the irregular warfare cards, which is possible by the nature of principle, which is singular. On this platform humanity writes itself a ticket to universal liberty and a new renaissance with such power that it can meet the Ice Age challenge on the platform of the strategic defence of humanity that then becomes possible.

The process of humanity writing itself a ticket

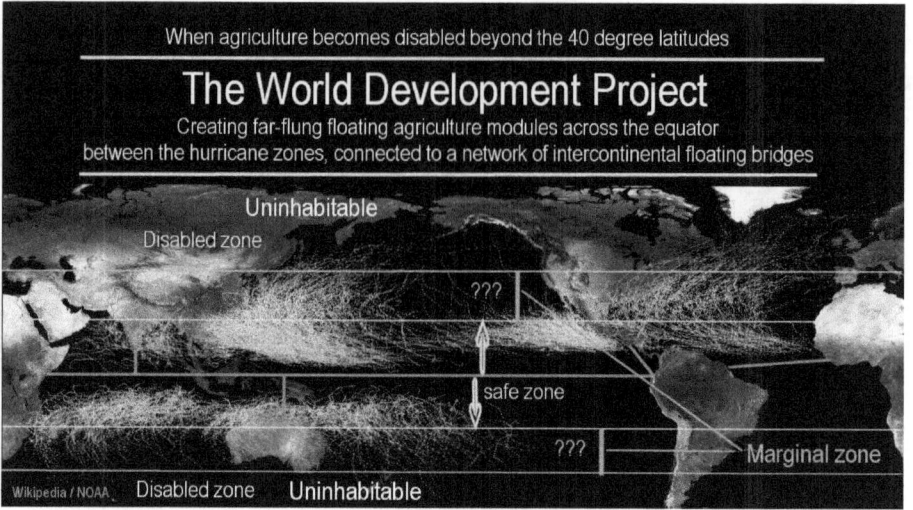

All this is possible, but will it be done? The process of humanity writing itself a ticket to meet the coming near Ice Age with a song, requires an immediate commitment, followed by the building of vast industries for the building of the vast infrastructures that are needed, followed by the building of the infrastructures themselves. That's the only ticket that can be written for humanity for a bright future in an Ice Age world. Today, for as far as I can see, the fields for developing this commitment are rather empty of people reaching for it. The scarecrow cards still rule the landscapes.

Historic Awaking

**#8 Historic Awaking

In the early days of the scarecrow cards

we want to be free

**Scientists Appealing
the Global Warming Doctrine**

The 1992 Heidelberg Appeal
The 1997 Leipzig Declaration
The Oregon Petition Project
Open letters to the Prime Minister of Canada
The 2007 U.S. Senate Report
100 prominent international scientists warn the U.N.
The new (2008) Oregon Petition Project (ongoing)

corel corp.

It was different in the early days. There was still some life left on the science scene. Large efforts were made in the early days of the scarecrow cards to keep the world mentally alive, by focusing on the known truth versus the known fraud. The movement started with the Heidelberg Appeal in 1992, that went out to the world from the university of Heidelberg in Germany, followed by the Leipzig Declaration, the Oregon Petition Project, and the huge 2007 U.S. Senate Report, all urging the world to wake up to an unfolding fraud against science and humanity. These were all movements towards the strategic defence principle.

The Heidelberg Appeal was supported by 4000 scientists from 69 countries, including nearly as many Nobel laureates. Of course their voices were not allowed to be heard at the Rio Earth Summit of the modern irregular war project.

The Leipzig Declaration
by 110 climate specialists

Reported in 1999 by Dr. Hugh Ellsaesser, an American atmospheric scientist associated with the Lawrence Livermore National Laboratory for 23 years, who also served 20 years as an Air Weather Officer for the U.S. Air Force.

Undeterred by the failure, the Leipzig Declaration was launched from the University of Leipzig. The project addressed only the narrow field of climatologists and climate specialists. The project collected 110 signatures from this highly specialized group. It was submitted to the Kyoto Climate Conference, and again it was not allowed to be heard.

In response to the Kyoto tragedy

In response to the Kyoto tragedy that resulted from the lack of truth in its premises, and had resulted in a global accord which the Russian Academy of Science termed "a global suicide pact," the really big Oregon Petition Project was launched. The petition project addressed all scientists worldwide in a campaign to urge the nations to reject the Kyoto accord. This time 17,000 people with an academic background responded. It may have been on the basis of this huge response that only a few countries actually ratified the accord.

The Senate collected 400 reports from individual respondents

The U.S. Senate Report

400 detailed individual declarations online

For details see: ice-age-ahead-iaa.ca

In 2007 the U.S. Senate report staged another scene of science opposition against the scare cards. The Senate didn't collect just signatures of agreement, but required written statements with details for the individuals' reasons for their stand. The Senate collected 400 of these reports from individual respondents. The reports are on file.

New Online Petition Project

31,487 mail-in responses
including submissions from 9,029 PhDs

It is often said to the present day that the carbon climate change theory is universally understood and is backed by a wide consensus among scientists. The statement is true only in the sense that it is widely understood in honest science that manmade carbon climate change is physically impossible. There appears to be a consensus in the community of real science that the promoted carbon scare theory is crap. Over 30,000 people with academic standing have made the effort to respond to a mail-in project as the latest effort. There appears to be a real consensus of standing in opposition, found in the science community.

The media created a consensus in guided dreaming

Of course the media also has created its own type of consensus that it is hired for, as is required for the modern irregular warfare objective. The media created a consensus in guided dreaming where nothing is fundamentally understood, where the theories are simply accepted and people bow to the degenerating effect they have. The result is that of a hidden type of terrorism.

The Ice Age challenge inspires the self-up-ramping of humanity

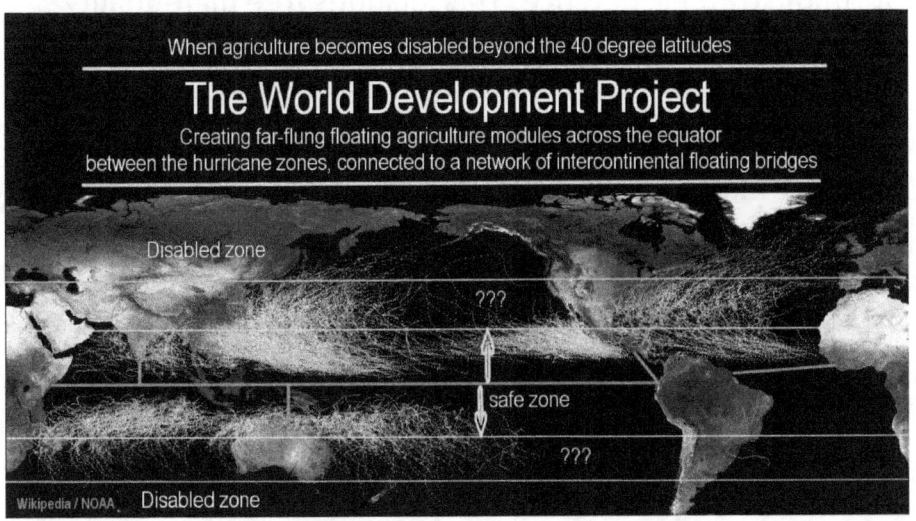

The Ice Age challenge, in contrast, has the opposite effect. It challenges society to truly awake, to ramp up its civilization. It inspires the self-up-ramping of humanity to ever-greater forms of economic development and scientific and technological progress. This is far from being a terrorist objective that puts society mentally into a coma and demands depopulation.

In real terms, the world needs more people, not less. The building of a new world in response to the Ice Age challenge that is now before us 30 years into the future, requires the combined effort of the entire world. It requires a full commitment to truth in science with a full commitment to the development of the power of the human mind, and the human potential.

This commitment on which civilization rests, amounts to

nothing less than the full commitment in society to the principle of the general welfare in all regards as a high-level form of strategic defence. This includes free high-quality universal housing, health care, education, with the remaining needs being met as freely as possible. The general welfare principle as a strategic defence against the poverty of small-minded thinking, is the minimal starting platform to spearhead the fullest development possible of the greatest asset that a society has, which is itself. If anyone thinks that the Ice Age challenge can be met on any lesser platform, this person is dreaming, and dreaming won't do.

Moments of Truth

**#9 Moments of Truth

The great German poet of freedom, Friedrich Schiller

Friedrich Schiller (1759–1805), Ölgemälde von Ludovike Simanowiz 1793/94

The great German poet of freedom, Friedrich Schiller had asked in his Aesthetical Letters, referring to the failed French Revolution, "Why is it that a great moment found such a little people?" He suggested that while the moment for great change in the flow of history had been at hand at the time, the moral condition in society to grasp the potential of the moment and to move with it, had been lacking. As many people had later realized with tears in the aftermath of the failure of the revolution, that the conditioning that they had lacked at the critical moment, became their lacking conditioning to survive on the personal level.

The great moment in our time

The great moment in our time is the Ice Age decision moment, and it will always be that until the Ice Age challenge is met.

The 2015 U.N. Climate Conference in Paris

Paris

Host city of the 2015 United Nations
Climate Change Conference,
COP 21 or CMP 11

The 2015 U.N. Climate Conference in Paris has the potential to serve as a stage for a moment of truth - for a moment of profound realization that stops the carbon dreaming; a moment to lay down the climate scarecrows; a moment for humanity to stand as giants in the defence of its future, and thereby of itself by getting series in considering the Ice Age Challenge.

Humanity had once taken a stand as giants

The treaty of Westphalia - a treaty by society with itself

1648

Ratification of the Peace of Münster between Spain and the Dutch Republic in the town hall of Münster, 15 May 1648. Painting by Gerard ter Borch the Younger (1617–1681)

Humanity had once taken such a stand as giants, and had accomplished what had never been accomplished before or has been since. It stood as giants, laterally, standing side by side. It ended nearly 100 years of war in 1648, not with a sword, but with a pen in hand and an open and alert mind. It created a peace that had rendered all humanity sovereign in principle and honoured with the commitment to promote what is of the advantage of the other, which is to everyone's advantage.

The meeting of giants in 1648 had been convened against the background of Johannes Kepler staging of a great moment for truth in the science of astrophysics that had liberated astronomy from centuries of dreaming. The pioneers in 1648 had latched onto this moment. It might have been their conditioning to become giants and stand tall, to stand as human beings.

The principle of universal sovereignty

The principle of universal sovereignty, which had been put onto the map in a big way in 1648 and was reflected in advancing science and technology thereafter, still stands today as the guiding star in civilization, though this star is fast being eroded and is almost universally denied in the fields of modern irregular warfare. The tower of Paris that you see before you, has been visited by more than 250 million people. It stands today as a lone reminder of a once proud past. The Paris U.N. Climate conference lacks the historic link to the critical moment of truth, as did all the preceding climate conferences of the last 40 years.

Where will we find the Johannes Kepler's of today

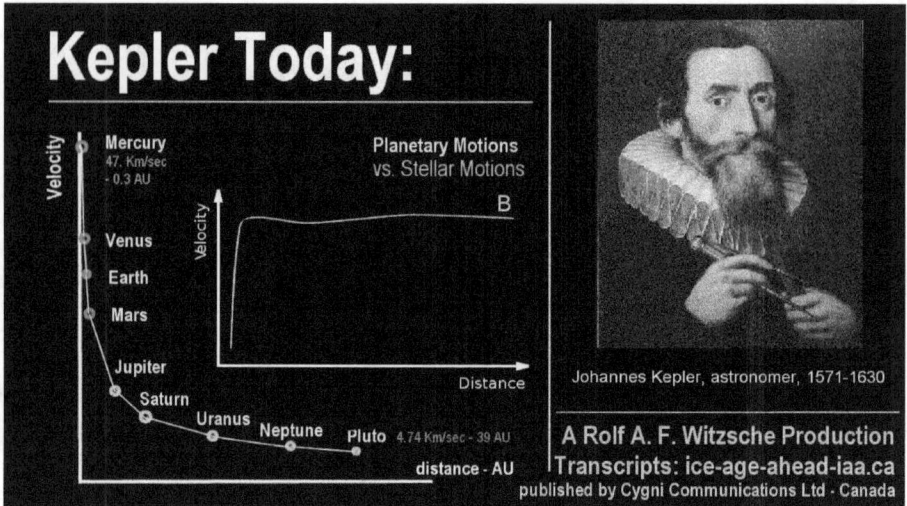

Where will we find the Johannes Keplers of today, like he who raised the standard of truth in astrophysics and set the stage for the giants to emerge on it?

Will the new song be sung that marks the end of the dreamtime?

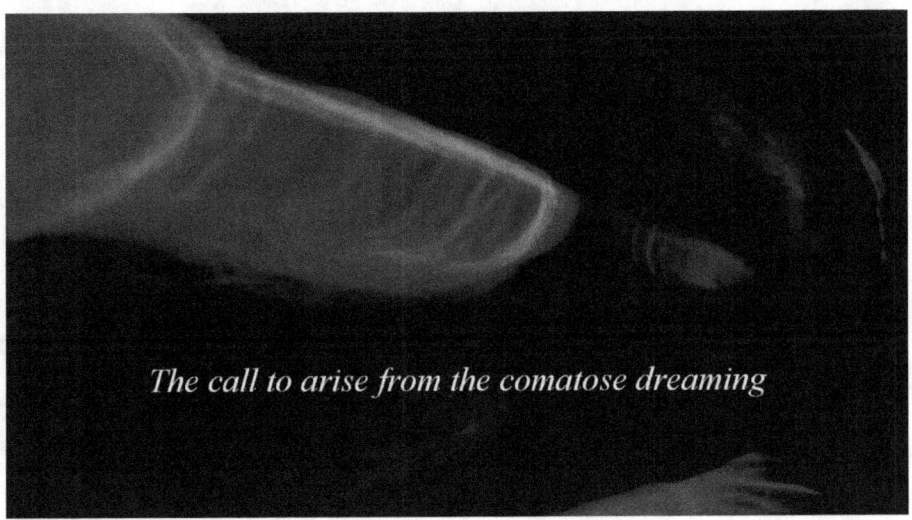

The call to arise from the comatose dreaming

It appears that this time, the call to arise from the comatose dreaming, will have to be powered by society itself. Society must make that call, which a human being should most certainly be able to make. No one can release society of this responsibility to itself.

Perhaps when this is understood, and perhaps only then when society does make this call to awake itself, will the new song be sung that marks the end of the dreamtime. It will be a simple, but profound song:

Oh dreamer leave behind your dream in joyful waking,

Oh dreamer leave behind your dream in joyful waking,
Oh captive rise, and 'wield' your power to be free,
Then stand on the hilltops, the golden truth proclaiming,
And paint the sky with the colors of universal liberty.

image (croped) courtesy of loriedarlin.tumblr.com

Oh dreamer leave behind your dream in joyful waking,
Oh captive rise, and 'wield' your power to be free,
Then stand on the hilltops, the golden truth proclaiming,
And paint the sky with the colors of universal liberty.

www.ingramcontent.com/pod-product-compliance
Lightning Source LLC
Chambersburg PA
CBHW070239190526
45169CB00001B/232